U.S. Marine Corps Close Combat

Plus

Rappelling

Department of Defense

Pentagon
Publishing

DEPARTMENT OF THE NAVY
Headquarters United States Marine Corps
Washington, D.C. 20380-1775

18 February 1999

FOREWORD

1. PURPOSE

Today's Marines operate within a continuum of force where conflict may change from low intensity to high intensity over a matter of hours. Marines are also engaged in many military operations other than war, such as peacekeeping missions or noncombatant evacuation operations, where deadly force may not be authorized. During non-combative engagements, Marines must determine if a situation warrants applying deadly force. Sometimes Marines must decide in a matter of seconds because their lives or the lives of others depend on their actions. To make the right decision, Marines must understand both the lethal and nonlethal close combat techniques needed to handle the situation responsibly without escalating the violence unnecessarily. Marine Corps Reference Publication (MCRP) 3-02B, *Close Combat,* provides the tactics, techniques, and procedures of Marine Corps close combat. It also provides the doctrinal basis for the Marine Corps Close Combat Training Program (MCCCTP).

2. SCOPE

This publication guides individual Marines, unit leaders, and close combat instructors in the proper tactics, techniques, and procedures for close combat. MCRP 3-02B is not intended to replace supervision by appropriate unit leaders and close combat instruction by qualified instructors. Its role is to ensure standardization and execution of tactics, techniques, and procedures throughout the Marine Corps.

3. SUPERSESSION

MCRP 3-02B supersedes Fleet Marine Force Manual (FMFM) 0-7, *Close Combat,* dated 9 July 1993. There are significant differences between the two publications. MCRP 3-02B should be reviewed in its entirety.

4. WARNING

Techniques described in this publication can cause serious injury or death. Practical application in the training of these techniques will be conducted in strict accordance with approved Entry Level Close Combat, Close Combat Instructor (CCI), and Close Combat Instructor Trainer (CCIT) lesson plans. Where serious danger exists, the reac er is alerted by the following:

WARNING

5. CERTIFICATION

Reviewed and approved this date.

BY DIRECTION OF THE COMMANDANT OF THE MARINE CORPS

J. E. RHODES
Lieutenant General, U.S. Marine Corps
Commanding General
Marine Corps Combat Development Command

DISTRIBUTION: 14400006600

OVERVIEW OF CLOSE COMBAT

1. Purpose of Close Combat

Close combat is the physical confrontation between two or more opponents. It involves armed and unarmed and lethal and nonlethal fighting techniques that range from enforced compliance to deadly force. The purpose of close combat is to execute armed and unarmed techniques to produce both lethal and nonlethal results. Unarmed techniques include hand-to-hand combat and defense against hand-held weapons. Armed techniques include techniques applied with a rifle, bayonet, knife, baton, or any weapon of opportunity.

2. Continuum of Force

Marines will find themselves in both combative and noncombative situations. The threat level in these situations can rise and fall several times based on the actions of both Marines and the people involved. The escalation of force stops when one person complies with the demands imposed by another person. This range of actions is known as a continuum of force. Continuum of force is the concept that there is a wide range of possible actions, ranging from voice commands to application of deadly force, that may be used to gain and maintain control of a potentially dangerous situation (MCO 5500.6_, *Arming of Security and Law Enforcement [EE] Personnel and the Use of Force).* Continuum of force consists of five levels that correspond to the behavior of the people involved and the actions Marines use to handle the situation (see the table below). Close combat techniques are executed in levels three, four, and five.

Level One: Compliant(Cooperative)

The subject complies with verbal commands. Close combat techniques do not apply.

Level Two: Resistant (Passive)

The subject resists verbal commands but complies immediately to any contact controls. Close combat techniques do not apply.

Continuum of Force		
Level	**Description**	**Actions**
1	Compliant (Cooperative)	Verbal commands
2	Resistant (Passive)	Contact controls
3	Resistant (Active)	Compliance techniques
4	Assaultive (Bodily Harm)	Defensive tactics
5	Assaultive (Serious Bodily Harm/ Death)	Deadly force
Note: Shading indicates levels in which Marines use close combat techniques.		

Level Three: Resistant (Active)

The subject initially demonstrates physical resistance. Marines use compliance techniques to control the situation. Level three incorporates close combat techniques to physically force a subject to comply. Techniques include—

- Come-along holds.
- Soft-handed stunning blows.
- Pain compliance through joint manipulation and the use of pressure points.

Level Four: Assaultive (Bodily Harm)

The subject may physically attack Marines, but he does not use a weapon. Marines use defensive tactics to neutralize the threat. Defensive tactics include the following close combat techniques:

- Blocks.
- Strikes.
- Kicks.
- Enhanced pain compliance procedures.
- Nightstick blocks and blows.

Level Five: Assaultive (Serious Bodily Harm/Death)

The subject usually has a weapon and will either kill or seriously injure someone if he is not stopped immediately and brought under control. Typically, to control the subject, Marines apply deadly force through the use of a firearm, but they may also use armed and unarmed close combat techniques.

3. Marine Corps Tactical Concepts

Close combat techniques support the following key Marine Corps tactical concepts. The concepts are not standalone ideas but are to be combined to achieve an effect that is greater than their separate sum.

Achieving a Decision

Achieving a decision is important in close combat. An indecisive fight wastes energy and possibly Marines' lives. Whether the intent is to control an opponent through restraint or defend themselves in war, Marines must have a clear purpose before engaging in close combat and act decisively once engaged.

Gaining an Advantage

A basic principle of martial arts is to use the opponent's strength and momentum against him to gain more leverage than one's own muscles alone can generate, thereby gaining an advantage. In close combat, Marines must exploit every advantage over an opponent to ensure a successful outcome. This can include employing various weapons and close combat techniques that will present a dilemma to an opponent. Achieving surprise can also greatly increase leverage. Marines try to achieve surprise through deception, stealth, and ambiguity.

Speed

Marines use speed to gain the initiative and advantage over the enemy. In close combat, the speed and violence of the attack against an opponent provides Marines with a distinct advantage. Marines must know and understand the basics of close combat so they can act instinctively with speed to execute close combat techniques.

Adapting

Close combat can be characterized by friction, uncertainty, disorder, and rapid change. Each situation is a unique combination of shifting factors that cannot be controlled with precision or certainty. For example, a crowd control mission may call for Marines to employ various techniques ranging from nonlethal restraint to more forceful applications. Marines who adapt quickly will have a significant advantage.

Exploiting Success

Typically, an enemy will not normally surrender simply because he was placed at a disadvantage. Marines cannot be satisfied with gaining an advantage in a close combat situation. They must exploit any advantage aggressively and ruthlessly until an opportunity arises to completely dominate the opponent. Marines must exploit success by using every advantage that can be gained.

(reverse blank)

Close Combat

Table of Contents

Page

Chapter 5. Throws

Chapter 6. Chokes and Holds

Chapter 7. Ground Fighting

Chapter 8. Nonlethal Techniques

Appendices

CHAPTER 1

FUNDAMENTALS OF CLOSE COMBAT

This chapter describes all techniques for a right-handed person. However, all techniques can be executed from eittwr side.

The Marine is depicted in camouflage utilities. The opponent is depicted without camouflage.

The fundamentals of close combat include ranges, weapons of the body, target areas of the body, and pressure points of the body. These fundamentals form the basis for all close combat techniques. They provide Marines with a common framework regardless of the type of confrontation or the techniques used. If Marines apply these fundamentals properly in a close combat situation, they may save their lives or the lives of fellow Marines.

1. Ranges of Close Combat

Close combat engagements occur within three ranges: long range, midrange, and close range. During any engagement, these ranges may blur together or may rapidly transition from one to another until either the opponent is defeated or the conflict is resolved.

Long Range

During long range engagements, combatants engage each other with rifles, bayonets, sticks, or entrenching tools. See figure below.

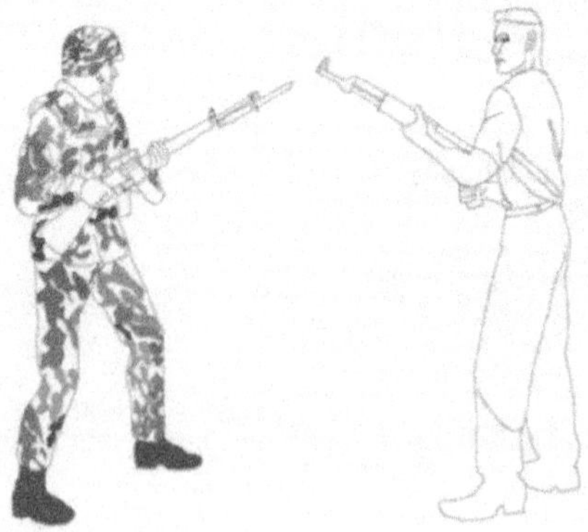

Midrange

During midrange engagements, combatants engage each other with knives, punches, or kicks.

Close Range

During close range engagements, combatants grab each other. Close range engagements also involve elbow strikes, knee strikes, and grappling.

2. Weapons of the Body

Hands and Arms

The hands, forearms, and elbows are the arm's individual weapons. The hands consist of several areas that can be used as weapons: fists, edges of hands, palms, and fingers.

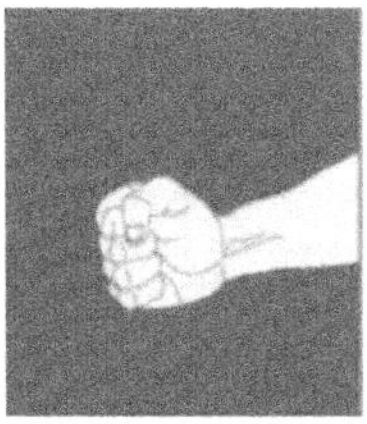

Fists. To minimize injury to the fists, Marines use their fists as weapons to target soft tissue areas such as the throat. The fists' striking surfaces are the first two knuckles of the hands or the meaty portions of the hands below the little fingers.

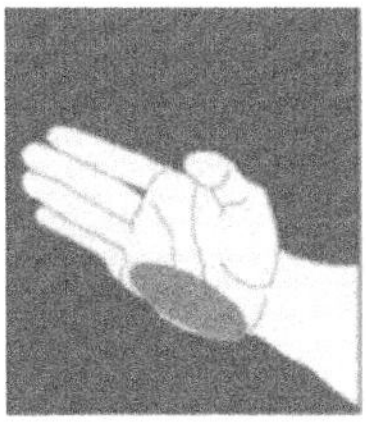

Edge of Hand. Marines use the edge of the hand (knife edge) as a weapon. Marines use the edge of the hand to strike soft tissue areas.

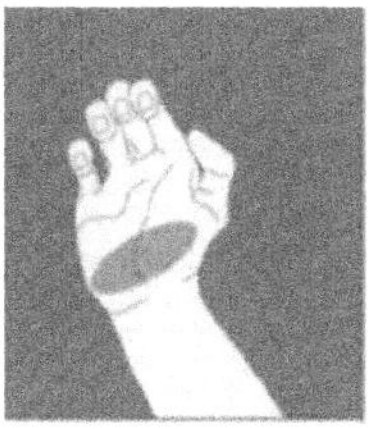

Palms. Because of the palm's padding, Marines use the heels of the palms to strike, parry, and/or block.

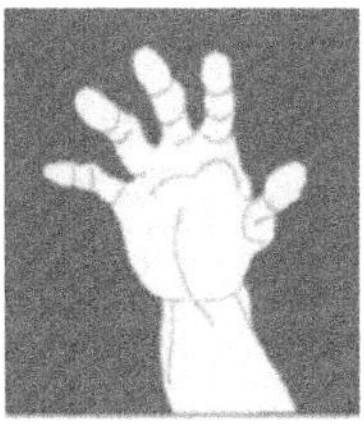

Fingers. Marines use the fingers to gouge, rip, and tear soft tissue areas (e.g., eyes, throat, groin).

Forearms. Marines use the forearms as a defensive tool to deflect or block attacks. Forearms can also be used as striking weapons to damage or break an opponent's joints and limbs. Marines sustain less self-injury when strikes are conducted with the forearms than when strikes are conducted with fists and fingers.

Elbows. Marines use the elbows as striking weapons. Because of the short distance needed to generate power, elbows are excellent weapons for striking during the close range of close combat.

Legs

The legs are more powerful than any other weapon of the body, and they are less prone to injury when striking. The feet are protected by boots and are the preferred choice for striking.

Feet. Marines use the balls of the feet, the insteps, and the toes to kick an opponent. Marines use the cutting edge of the heels and the heels to stomp on an opponent. Marines must be wearing boots when striking with the toes.

Knees. Like elbows, knees are excellent weapons in the close range of close combat. Knee strikes are most effective while fighting close to an opponent where kicks arc impractical. The opponent's groin area is an ideal target for the knee strike if he is standing upright. Knee strikes can deliver a devastating secondary attack to an opponent's face following an initial attack that caused him to bend at the waist.

3. Target Areas of the Body

During close combat, Marines strive to attack the accessible target areas of an opponent's body. The readily accessible areas will vary with each situation and throughout the engagement. The target areas are divided into five major groups: head, neck, torso, groin, and extremities. The figure below illustrates target areas of the body.

Head

The vulnerable regions of the head are the eyes, temple, nose, ears, and jaw. Massive damage to the head kills an opponent.

Eyes. The eyes are excellent targets because they are soft tissue areas that arc not protected by bone or muscle. Attacks to this area may cause the opponent to protect the area with his hands, allowing Marines to execute a secondary attack to other

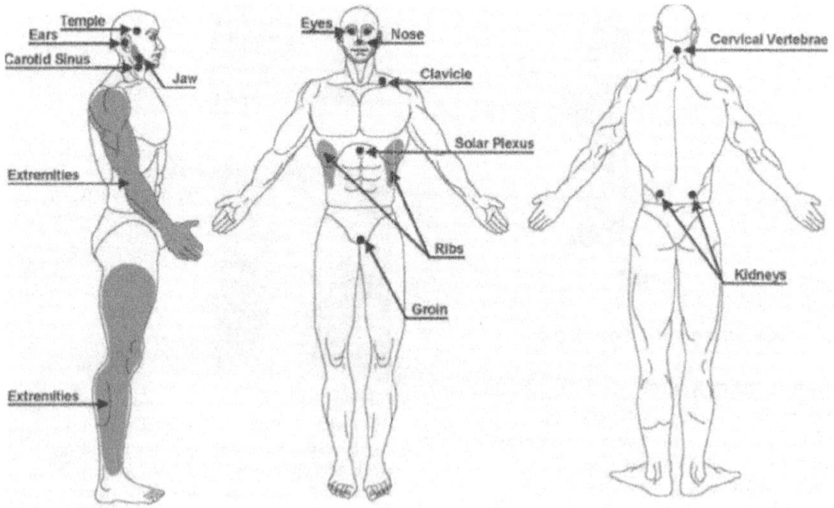

target areas while the opponent uses his hands to protect his eyes.

Temple. The temple is one of the most fragile areas of the head. Powerful strikes to the opponent's temple cause permanent damage and death.

Nose. The nose is very sensitive and easily broken. An attack to this area causes involuntary watering and closing of the opponent's eyes, rendering him vulnerable to secondary attacks. However, through training, individuals can condition themselves to withstand attacks to the nose. Therefore, any attack to the nose must be powerfully delivered.

Ears. Attacks to the ears may cause the eardrum to rupture. But this may not stop or even distract an opponent unless Marines powerfully deliver the strike.

Jaw. The jaw region, when struck forcefully, renders the opponent unconscious. Strikes to the jaw cause painful injuries to the teeth and surrounding tissues (e.g., lips, tongue), but the risk of self-injury is great unless Marines deliver strikes with a hard object such as a helmet, rifle butt, or boot heel.

Neck

The front of the neck, or throat area, is a soft tissue area that is not covered by natural protection. Damage to this region causes the opponent's trachea to swell, closing his airway, which can lead to death.

Carotid Sinus. The carotid sinus is located on both sides of the neck just below the jaw. Strikes to the carotid sinus restrict blood flow to the brain, causing loss of consciousness or death.

Cervical Vertebrae. The cervical vertebrae on the back of the neck, from the base of the skull to the top of the shoulders, contains the spinal cord, which is the nervous system's link to the brain. The weight of the head and the lack of large muscle mass allow damage to the cervical vertebrae and spinal cord. Excessive damage to this area causes pain, paralysis, or death.

Torso

Clavicle. The opponent's clavicle (or collar bone) can be easily fractured, causing immobilization of the arm.

Solar Plexus. Attacks to the opponent's solar plexus or center of the chest can knock the breath out of him and immobilize him.

Ribs. Damage to the opponent's ribs immobilizes him. It may also cause internal trauma.

Kidneys. Powerful attacks to the opponent's kidneys cause immobilization, permanent damage, or death.

Groin

The groin area is another soft tissue area not covered by natural protection. Any damage to this area causes the opponent to involuntarily protect his injured area, usually with his hands or legs. In male opponents, the scrotum is the main target since even a near miss causes severe pain, contraction of the lower abdominal muscles, deterioration of his stance, and possible internal trauma.

Extremities

Rarely will an attack to the opponent's extremities (arms and legs) cause death, but they are important target areas in close combat. Damage to an opponent's joints causes immobilization.

4. Pressure Points of the Body

There are nerves in the human body that, when pressure is applied or when they are struck, allow Marines to control a subject through pain compliance. Marines use pressure points to control an opponent when deadly force is not authorized. They also use pressure points to soften or distract an opponent so a lethal or nonlethal technique can be employed. The figure on page 1-5 illustrates the body's pressure points. Marines execute attacks to pressure points by—

- Rapidly kicking or striking pressure points.
- Slowly applying steady pressure to pressure points.

Infraorbital Nerve

The infraorbital nerve is just below the nose. Marines apply pressure to this nerve with an index finger to control the opponent.

Mastoid Process

The mastoid process is behind the base of the ear and beneath the edge of the jaw. Marines apply inward and upward pressure to this pressure point with the fingers to distract and control the opponent.

Jugular Notch

The jugular notch is at the base of the neck in the notch formed at the center of the clavicle. Marines apply pressure in a quick, stabbing motion with the index finger. Strikes to the jugular notch cause serious damage.

Brachial Plexus (Tie In)

The brachial plexus (tie in) is on the front of the shoulder at the joint. Strikes and pressure applied with the hand are effective on this nerve.

Radial Nerves

Radial nerves are on the inside of the forearms along the radius bones. Strikes and pressure applied with the hand to the radial nerve serve as a softening technique.

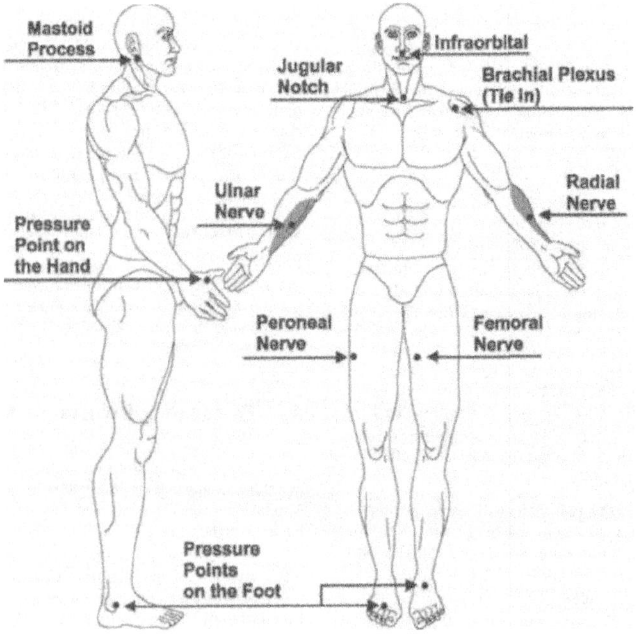

Ulnar Nerve

Ulnar nerves are on the outside of the forearms along the ulnar bones. Strikes and pressure applied with the hand to the ulnar nerve serve as a softening technique.

Pressure Point on the Hand

The hands contain a pressure point on the webbing between the thumbs and index fingers where the two bones of the fingers meet.

To force an opponent to soften or release his grip, Marines apply pressure with their index fingers to this pressure point or strike this pressure point with their fists.

Femoral Nerves

Femoral nerves are on the inside of the thighs along the femur bones. Strikes to the femoral nerve serve as a softening technique.

Peroneal Nerves

Peroneal nerves are on the outside of the thighs along the femur bones. Strikes to the peroneal nerve serve as a softening technique.

Pressure Points on the Feet

There are pressure points on the feet that, when pressure is applied or when they are struck, serve to soften or distract the opponent. Marines apply pressure with the toe, edge, or heel of their boots to the following points:

- The notch below the ball of the ankle.
- The top center of the foot, above the toes.
- The top of the foot where the leg and foot meet.

5. Basic Warrior Stance

Achieving a solid stance is critical to stability and movement throughout any close combat confrontation. The basic warrior stance provides the foundation for stability and movement that is needed to execute close combat techniques. To execute the basic warrior stance, Marines put their feet apart, hands up, elbows in, and chin down.

Feet Apart

Place feet shoulder-width apart.

Keep the head forward and eyes on the opponent, take a half step forward with the left foot, and pivot on the heels so the hips and shoulders are at approximately a 45-degree angle to the right.

Distribute body weight evenly on both legs. Bend the knees slightly.

Hands Up

Curl the fingers naturally into the palm of the hand. Position the thumb across the index and middle fingers. Do not clench the fists. Clenching the fists increases muscular tension in the forearms and decreases speed and reaction time.

Bring the hands up to the face at chin level, with the palms facing each other. Hold the fists up high enough to protect the head, but not so high that they block the field of vision. Ensure continuous eye contact with the opponent.

Elbows In

Tuck the elbows in close to protect the body.

Chin Down

Tuck the chin down to take advantage of the natural protection provided by the shoulders.

6. Angles of Approach and Movement

Marines use movement to control a confrontation and to retain a tactical advantage. Movement increases power and maximizes momentum. By moving around the opponent, Marines gain access to different target areas of the opponent's body. Movement allows Marines to use different weapons of the body and different close combat techniques to attack specific target areas.

Angles of Approach

Marines move anywhere within a 360-degree circle around the opponent to gain a tactical advantage. This circle provides access to different target areas of the opponent's body.

When facing an opponent, Marines move in a 45-degree angle to either side of the opponent. Moving at a 45-degree angle avoids an opponent's strike and puts Marines in the best position to attack the opponent. Marines should avoid being directly in front of an opponent during a confrontation. If a Marine is directly in front of an opponent, the opponent can rely on his forward momentum and linear power to seize the tactical advantage.

Movement

Marines must know how to move in all directions while maintaining the basic warrior stance. During any movement, the legs or feet should not be crossed. Once a movement is completed, the basic warrior stance should be resumed. Maintaining the basic warrior stance protects Marines and puts them in the proper position to launch an attack against an opponent.

Note: Before body movement begins. Marines turn their heads quickly to the new direction. The faster the head turns, the faster the body moves, and the quicker Marines attain visual contact with the opponent.

Forward to the Left. To move forward to the left from the basic warrior stance, Marines-

: Move the left foot forward at a 45-degree angle from the body (approximately 12 to 15 inches), keeping the toe pointed toward the opponent.

Bring the right foot behind the left foot as soon as the left foot is in place. This returns Marines to the basic warrior stance.

Forward to the Right. To move forward to the right from the basic warrior stance, Marines-

Move the right foot forward at a 45-degree angle from the body (approximately 12 to 15 inches).

Bring the left foot, toe pointing toward the opponent, in front of the right foot as soon as the right foot is in place. This returns Marines to the basic warrior stance.

Backward to the Left. To move backward to the left from the basic warrior stance, Marines execute the forward movement in reverse. Marines-

Move the left foot backward at a 45-degree angle from the body (approximately 12 to 15 inches), keeping the toe pointed toward the opponent.

Bring the right foot behind the left foot as soon as the left foot is in place. This returns Marines to the basic warrior stance.

Backward to the Right. To move backward to the right from the basic warrior stance, Marines execute the forward movement in reverse. Marines-

Move the right foot backward at a 45-degree angle from the body (approximately 12 to 15 inches).

Bring the left foot, toe pointing toward the opponent, in front of the right foot as soon as the right foot is in place. This returns Marines to the basic warrior stance.

7. Balance and Off-Balancing

Balance

In any close combat situation, Marines must strive to maintain balance. The last place to be in a close combat situation is on the ground. Marines must maintain a strong base and a low center of balance, their feet must be a shoulder-width apart, and they must stay on their toes to enable quick movement.

Off-balancing

Marines use off-balancing techniques to control an opponent. These techniques are used to throw an opponent to the ground while Marines remain standing, or they are used to put Marines in a position for an offensive attack.

Off-balancing techniques use the opponent's momentum to move or throw him. For example, if the opponent is charging a Marine, the Marine pulls the opponent to drive him to the ground. Likewise, if the opponent is pulling a Marine, the Marine pushes the opponent to drive him to the ground.

Off-balancing techniques also rely on the power generated by the opponent. For example, during combat a Marine may be tired or outnumbered. Depending on the generated energy and momentum of the opponent, the Marine employs off-balancing techniques with very little effort and still provides effective results.

Because off-balancing techniques rely on the momentum and power generated by the opponent, these techniques are particularly effective for Marines who may be outsized by their opponent or lack their opponent's strength.

Angles of Off-balancing

There are eight angles or directions in which an opponent can be off-balanced: forward, rear, right, left, forward right, forward left, rear right, and rear left.

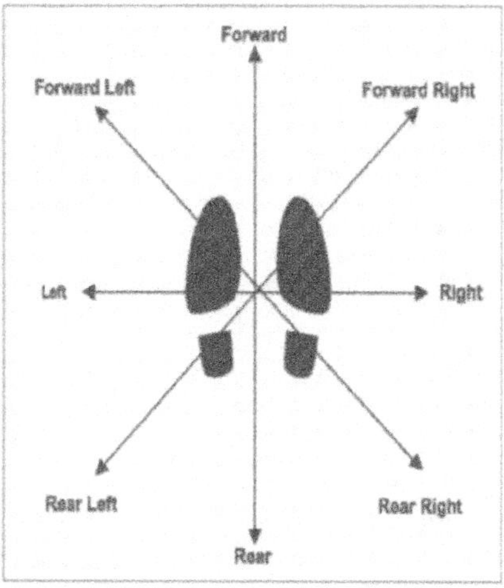

Note: The angles correspond to the Marine's perspective, not the opponent's. Forward, rear, right, and left are straight angles. Forward right, forward left, rear right, and rear left are considered quadrants that are at a 45-degree angle in either direction to the front or rear.

Off-balancing Techniques

Marines off-balance an opponent by pushing, pulling, or bumping the opponent with their hands, arms, or bodies.

To pull, Marines grab an opponent with their hands and drive him forcefully to one of the rear quadrants or to the right or left.

To push, Marines grab the opponent with their hands and drive him forcefully into one of the front quadrants or to the right or left.

Marines execute bumping in the same manner as pushing, but use their shoulders, hips, and legs instead of their hands to off-balance the opponent.

8. Falls

Marines may lose their balance or be thrown to the ground during encounters with an opponent. Marines use falling techniques to absorb the impact of a fall and to quickly return to their feet following an opponent's attack.

Whether falling or being thrown by an opponent, Marines strive to reduce the force of the impact, to prevent serious personal injury, and to increase the chances of survival. Falling techniques use the body's large muscles (back, thighs, buttocks) to protect vital organs and bones from injury and immobilization.

Front Fall

Marines execute a front fall to break a fall on the front. To execute the front fall, Marines-

Bend the elbows and place the palms facing out in a position to spread and absorb the impact of the fall.

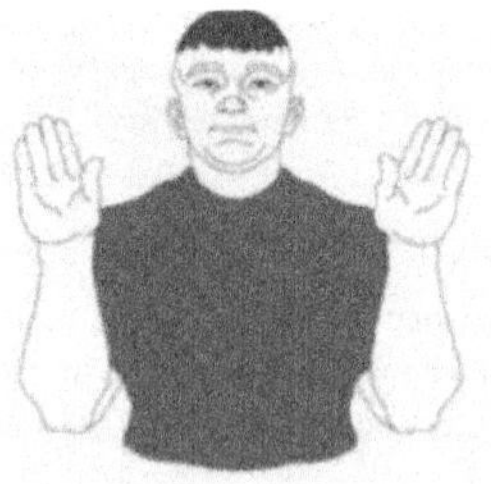

Fall forward, breaking the fall with the forearms and palms. The forearms and hands, down to the fingertips, should strike the ground simultaneous-iy-

Offer resistance with the forearms and hands to keep the head raised off the ground.

Side Fall

Marines execute a side fall to break a fall on the side. To execute the side fall, Marines-

Bring the right arm across the body so the hand is next to the left shoulder with the palm facing inboard.

Fall to the side, breaking the fall with the right arm by slapping the ground and making contact from the shoulder or forearm down to the hand. At the same time, tuck the chin and keep the head raised off the ground. The chin should be tucked to the chest at all times to prevent whiplash.

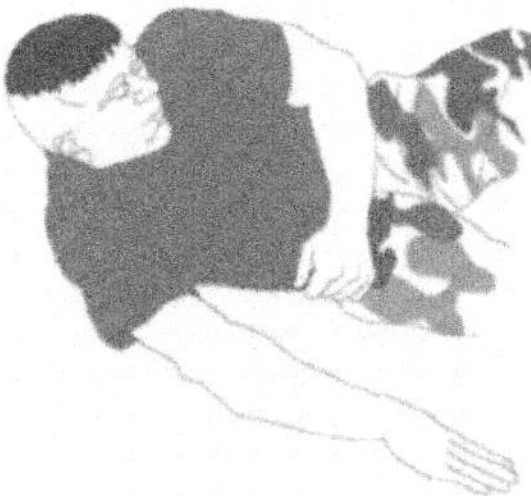

Stretch the right leg out to make contact with the ground and to distribute and absorb the impact. Bend the left leg, allowing the foot to make contact with the ground.

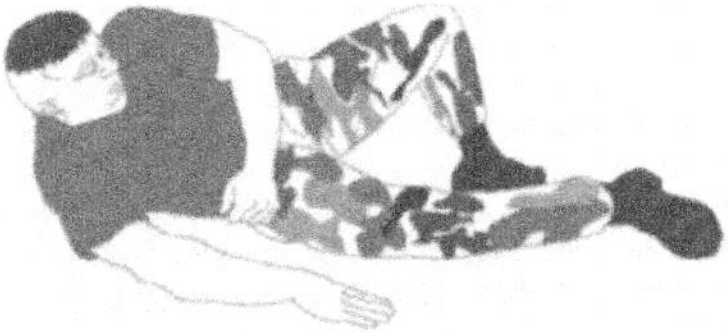

Back Fall

Marines execute a back fall to break the fall when being thrown or falling backward. To execute the backfall, Marines-

Cross the hands in front of the chest and tuck the chin.

Fall backward and slap the ground with the forearms and hands to absorb the impact of the fall and keep the head off the ground.

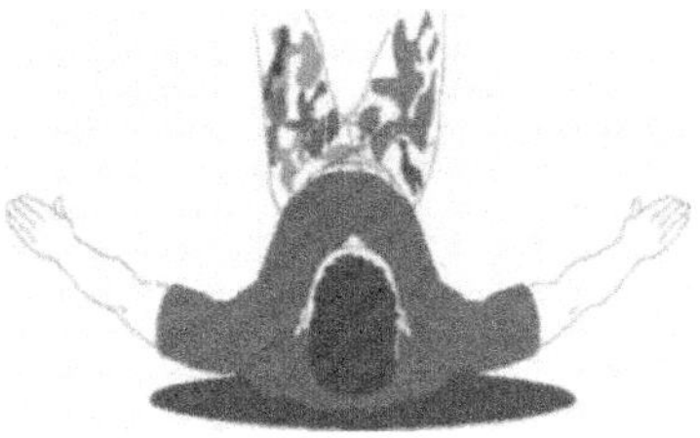

Forward Shoulder Roll

Marines use the forward shoulder roll to break a fall from an opponent's attack and to use the momentum of the fall to get back on their feet quickly. Ideally, Marines execute the forward shoulder roll to a standing position so they can continue fighting. To execute the forward shoulder roll, Marines-

Contact the ground with the back of the right forearm and upper arm. Tuck the chin into the chest.

Roll onto the right shoulder, rolling diagonally across the back to land on the left hip.

Slap the ground with the left arm, absorbing the impact from the shoulder to the hand, palm down.

Keep the left leg straight to absorb as much of the impact as possible. The right leg is bent and the foot hits flat on the ground.

Bend the left leg upon impact to push off with the left knee and leg to a squatting and then a standing position. Forward momentum should carry the Marine to a standing position.

(reverse blank)

CHAPTER 2

LETHAL AND NONLETHAL WEAPONS TECHNIQUES

This chapter describes all techniques for a right-handed person. However, all techniques can be executed from either side.

In drawings, the Marine is depicted in woodland camouflage utilities; the opponent is depicted without camouflage. In photographs, the Marine is depicted in woodland camouflage utilities; the opponent is depicted in desert camouflage utilities.

1. Bayonet Techniques

WARNING

During training, Marines must have bayonets sheathed. Marines use bayonet dummies to practice bayonet techniques. When practicing offensive and defensive bayonet techniques student-on-student, Marines use pugil sticks.

All Marines armed with a rifle carry a bayonet. The bayonet is an effective weapon if Marines are properly trained in offensive and defensive bayonet techniques. An offensive attack, such as a thrust, is a devastating attack that can quickly end a fight. Defensive techniques, such as the block and parry, can deter the opponent's attack and allow Marines to regain the initiative. Through proper training, Marines develop the courage and confidence required to effectively use a bayonet to protect themselves and destroy the enemy. In situations where friendly and enemy troops are closely mingled and rifle fire and grenades are impractical, the bayonet becomes the weapon of choice.

Holding the Rifle

To execute bayonet techniques, Marines hold the rifle in a modified basic warrior stance. All movement begins and ends with the basic warrior stance. To hold the rifle, Marines—

Use an overhanded grasp to grab the small of the rifle's stock. Use an underhanded grasp to grab the handguards of the rifle.

Lock the buttstock of the rifle against the hip with the right forearm.

Orient the blade end of the rifle toward the opponent.

Offensive Bayonet Techniques

Straight Thrust. Marines use the straight thrust to disable or kill an opponent. It is the most deadly offensive technique because it causes the most trauma to an opponent. Target areas are the opponent' s throat, groin, or face. The opponent's chest and stomach are also excellent target areas if not protected by body armor or combat equipment. To execute the straight thrust, Marines—

Lift the left leg and lunge forward off the ball of the right foot while thrusting the blade end of the weapon forward, directly toward the opponent.

Retract the weapon and return to the basic warrior stance.

Slash. Marines use the slash to kill an opponent or to create an opening in his defense. The target area is the opponent's neck. To execute the slash, Marines—

Extend the left hand back toward the left shoulder.

Thrust the left hand forward and swing it to the right, bring the right hand back toward the hip, and turn the cutting edge of the blade toward the opponent's neck. The movement is a slashing motion so the blade cuts across the opponent's neck.

Horizontal Buttstroke. Marines use the horizontal buttstroke to weaken an opponent's defenses, to cause serious injury, or to set him up for a killing blow. Target areas are the opponent's head, neck, and legs. To execute the horizontal buttstroke, Marines—

Step forward with the right foot and drive the right hand forward. Rotate the hips and shoulders into the strike. Move the left hand back toward the left shoulder.

Strike the opponent with the butt of the weapon.

Vertical Buttstroke. Marines use the vertical buttstroke to weaken an opponent's defenses, to cause serious injury, or to set him up for a killing blow. Target areas are the opponent's groin and face. To execute the vertical buttstroke, Marines—

Step forward with the right foot and drive the right hand straight up.

Pull the left hand back over the left shoulder.

Strike the opponent with the butt of the weapon.

Smash. Marines use the smash as a follow-up technique to the horizontal or vertical buttstroke, particularly if they missed the target. The target area is the opponent's head. To execute the smash following a buttstroke, Marines-

Step forward with the right foot and place the blade end of the weapon over the left shoulder and elevate the right elbow above the shoulders.

Drive the arms straight forward, striking the opponent with the butt of the weapon.

Defensive Bayonet Techniques

Parry. Marines use a parry as a defensive technique to redirect or deflect an attack. A parry is a slight redirection of a linear attack by an opponent; e.g., a straight thrust or a smash. To execute the parry, Marines-

Use the bayonet end of the rifle to redirect the barrel or bayonet of the opponent's weapon.

Lock the weapon against the hip with the right forearm.

Rotate to the right or left, moving the bayonet end of the rifle to parry the opponent's attack. Rotation is generated from the hips.

Redirect or guide the opponent's weapon away from the body by exerting pressure against the opponent' s weapon.

High Block. Marines execute a high block against a vertical attack coming from high to low. To execute the high block, Marines—

Thrust the arms up forcefully at approximately a 45-degree angle from the body. The weapon should be over the top of the head and parallel to the ground. The elbows are bent, but there is enough muscular tension in the arms to absorb the impact and deter the attack.

Low Block. Marines execute the low block

against a vertical attack coming from low to high. To execute the low block, Marines—

Thrust the arms down forcefully at approximately a 45-degrce angle from the body. The weapon

should be at or below the waist and parallel to the ground. The elbows are bent, but there is enough muscular tension in the arms to absorb the impact and deter the attack.

Left and Right Block. Marines execute a left or right block against a horizontal buttstroke or a slash. To execute the left or right block, Marines-

Thrust the arms forcefully to the right or left, holding the rifle vertically in the direction of the attack. The elbows are bent, but there is enough muscular tension in the arms to absorb the impact and deter the attack.

Counter Action Following the Block. After deflecting an opponent's attack with a block, Marines counter with a slash or a horizontal buttstroke to regain the initiative. However, the objective in any bayonet fight is to thrust forward with the blade end of the weapon to immediately end the fight.

Group Strategy

On occasion, Marines may engage an opponent as a member of a group or numerous opponents by one's self or as a member of a group. By combining bayonet fighting movements and simple strategies, Marines can effectively overcome their opponent or opponents.

Offensive Strategy: Two Against One. If two bayonet fighters engage one opponent, the fighters advance together.

Fighter 1 engages the opponent while fighter 2 swiftly and aggressively attacks the opponent's exposed flank and destroys the opponent.

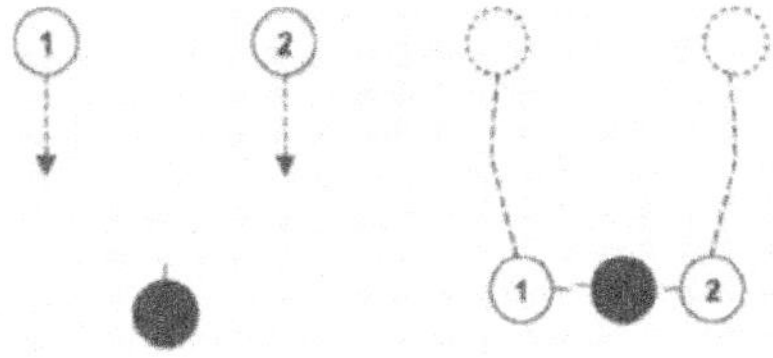

Offensive Strategy: Three Against Two. If three bayonet fighters engage two opponents, the fighters advance together keeping their opponents to the inside.

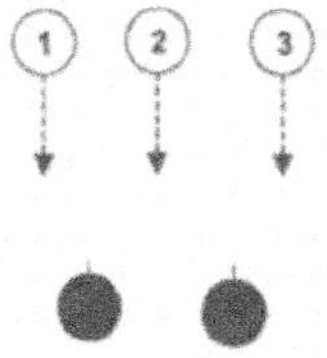

Fighters 1 and 3 engage opponents. Fighter 2 attacks the opponent's exposed flank engaged by fighter 1 and destroys the opponent.

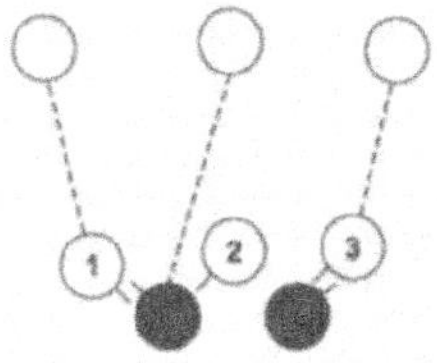

Fighters 1 and 2 turn and attack the exposed flank of the opponent engaged by fighter 3 and destroy the opponent.

Defensive Strategy: One Against Two. If a fighter is attacked by two opponents, the fighter immediately positions himself at the flank of the nearest opponent and keeps that opponent between himself and the other opponent.

Using the first opponent's body as a shield against the second opponent, the fighter destroys the first opponent quickly before the second opponent moves to assist.

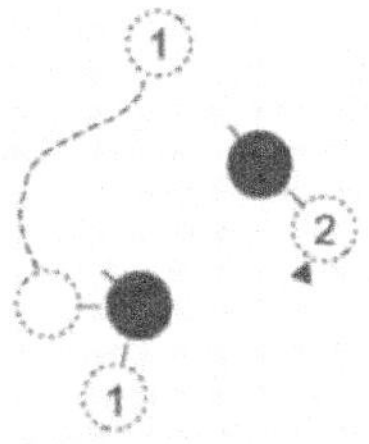

Then, the fighter engages and destroys the second opponent.

Defensive Strategy: Two Against Three. If two fighters are attacked by three opponents, the fighters immediately move to the opponent's flanks.

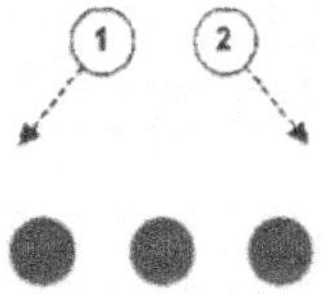

Fighters 1 and 2 quickly attack and destroy their opponents before the third opponent closes in.

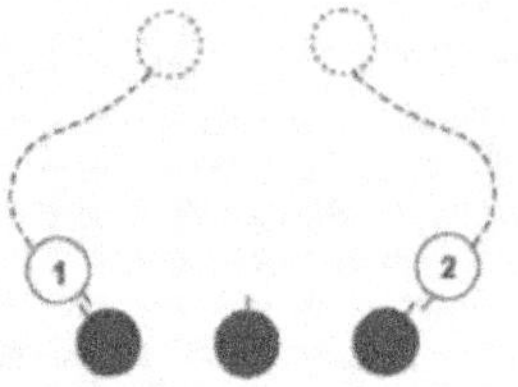

Fighter 1 engages the third opponent while fighter 2 attacks the opponent's exposed flank and destroys the opponent.

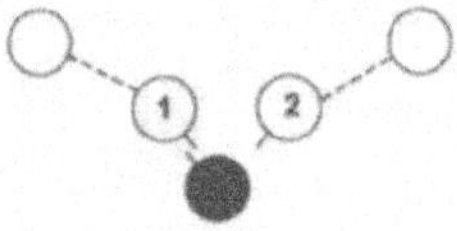

2. Nonlethal Rifle and Shotgun Retention Techniques

Most Marines are armed with the M16A2 service rifle. Marines are taught to keep their weapon with them at all times. Marines must be constantly alert to their surroundings and the people moving in and around their environment. Marines may be confronted by an individual who tries to take theii weapons. If this happens, Marines should not struggle with the individual. To retain positive control of their weapons, Marines must understand and apply weapons retention techniques, otherwise known as armed manipulation. The fol lowing techniques can be used with either the rifle or the shotgun.

Blocking Technique

To execute a blocking technique, Marines-

Stand in a defensive posture.

Use the weapon to block the opponent by thrusting it out firmly, with the elbows bent. Do not try to hit the opponent with the rifle. The rifle is used as a barrier.

Technique if Opponent Grabs Weapon Underhanded

Usually an opponent will try to grab the weapon or block it as an instinctive action. If the opponent uses an underhand grab to seize the handguards of the weapon, Marines—

Trap the opponent's closest finger(s) above the knuckle with the thumb so he cannot release his grip-

Apply bone pressure on the opponent's finger to initiate pain compliance.

Rotate the barrel of the weapon up or down quickly while maintaining pressure on the opponent's hand. At the same time, quickly pivot to off-balance the opponent.

Technique if Opponent Grabs Weapon Overhanded

If the opponent uses an overhand grab to seize the weapon overhanded, Marines—

Trap the opponent's finger to hold his hand in place.

Rotate the barrel to place it across the opponent's forearm and apply downward pressure. This action is similar to an armbar.

Technique if the Opponent Grabs the Muzzle

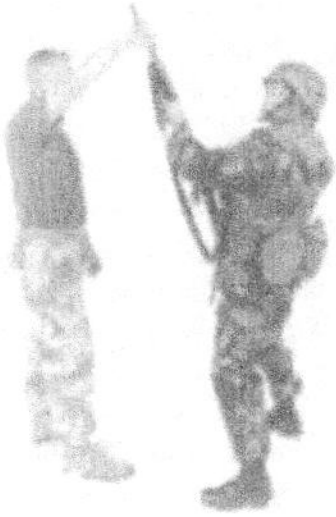

If the opponent grabs the muzzle of the weapon, Marines—

Rotate the muzzle quickly in a circle motion. Slash downward with the muzzle to release the opponent's grip.

Butt Strikes

Strikes with the butt of the weapon control or ward off an attacker. During any of the retention techniques, Marines use the heel or cutting edge of the weapon to deliver butt strikes to the inside or the outside of the opponent's thigh. The inside butt strike targets the femoral nerve as illustrated in the photo to the left.

The outside butt strike targets the peroneal nerve as illustrated in the photo to the left. Strikes can be made to the outside or inside of the thighs. If a strike to one side of the thigh misses, Marines follow back through with the butt of the weapon on the other side of the thigh.

Off-balancing Techniques

Marines apply off-balancing techniques to throw an opponent to the ground and retain possession of the weapon.

If the opponent grabs the weapon and pushes, Marines should not push on the weapon. They should—

Move with the momentum and movement of the opponent by pivoting in the direction of the movement by stepping back.

Throw the opponent to the ground with a quick jerking movement by lowering the mu/./le and swinging the butt of the weapon.

If the opponent grabs the weapon and pulls, Marines—

Step on the opponent's foot and push forward to off-balance him and drive him to the ground.

Sweep the opponent's feet out from under him by hooking his leg with the leg and kicking backward.

3. Nonlethal Handgun Retention Techniques

Many Marines are armed with the M9 service pistol. Marines must keep their weapons in their possession at all times. Marines must be constantly alert to their surroundings and the people moving in and around the environment. Marines may be confronted by an individual who tries to take their weapons. To retain positive control of the weapon, Marines must understand and apply handgun retention techniques.

Blocking Technique

Marines perform the following blocking techniques if an opponent attempts to grab their pistol in the holster. Marines—

Place the body between the weapon and the opponent by immediately pivoting so the weapon is away from the opponent.

Step back and away from the opponent while placing the hand on the pistol grip.

Extend the left hand and block, deflect, or strike the opponent's arm.

Armbar Technique

Marines use the armbar technique when an opponent uses his right hand to grab their pistol in the holster. To execute the armbar technique, Marines—

Trap the opponent's right hand by grasping the opponent's wrist or hand with the right hand and applying pressure against the body.

Step back with the right foot and pivot sharply to the right to be next to the opponent. Always pivot in a direction that keeps the weapon away from the opponent.

Straighten the opponent's arm to apply an armbar. The arm should be straight across the torso.

Continue pivoting to the right while pulling back on the opponent's shoulder. This action may break the opponent's arm.

Place the left hand across the opponent's face and apply pressure back and down to take the oppo-

nent to the ground. Pressure applied beneath the nose or on the trachea is equally effective.

Wristlock Technique

Marines use the wristlock technique when an opponent grabs their pistol while it is in the holster with his right hand. To execute the w ristlock technique, Marines—

Grasp the opponent's wrist or hand with the right hand and apply pressure against the body.

Step back and away from the opponent and pivot to the right so the weapon is away from the opponent. Always pivot in a direction that keeps the weapon away from the opponent.

Reach over the opponent's arm with the left hand and grab his hand, applying pressure against his arm with the left forearm. Execute a wristlock.

Incorporate the second hand into the wristlock and, stepping back with the left foot, pivot to the left.

Execute a two-handed wristlock by exerting downward pressure with the thumb and rotating his hand to the left.

Continue to pivot to off-balance the opponent and drive him to the ground.

Note: If the opponent grabs the pistol with his left hand, Marines execute the wristlock with one hand and step in toward the opponent, rather than

away from the opponent.

Softening Techniques

Handgun retention techniques use softening techniques applied to pressure points. Bone pressure

and strikes with the hands (i.e., hammer fist), knees, and feet are also effective softening techniques.

Pressure Points. Marines use pressure point techniques to get the opponent to loosen his grip. Marines use their finger tips to apply pressure to the webbing between the index finger and thumb, the jugular notch, and the brachial plexus tie in. The following figure illustrates pressure applied to the brachial plexus tie in.

Bone Pressure. Bone pressure is the application of pressure on a bone against a hard object to initiate pain compliance. To apply bone pressure, Marines use their hand to trap the opponent's hand on the weapon. Marines apply a slow, steady pressure to the opponent's hand and fingers until his grip is softened or he releases his hold.

Strikes. If it is difficult to apply a retention technique, Marines employ strikes or kicks to force the opponent to loosen his grip. Strikes to the

eyes, the arms (radial nerve), or shoulder (brachial plexus tie in) soften the opponent's grip on the weapon.

Kicks and knee strikes to the peroneal nerve, the femoral nerve, or the groin are effective because the opponent is typically unprepared to counter the strike.

Stomping on the top of the opponent's foot may distract him or loosen his grip on the weapon.

4. Firearm Disarmament Techniques

Marines use firearm disarmament techniques during a close-range confrontation if they are un-

armed and the opponent has a firearm (pistol). These techniques are equally effective if Marines are armed but do not have time to withdraw and present the weapon. The goal of firearm disarmament techniques is to gain control of the situation

so Marines gain the tactical advantage. The goal is not necessarily to get control of the opponent's weapon.

Pistol to the Front

This technique is used when Marines are unarmed and the opponent has a pistol pointing at their front (e.g., head, chest). The technique is the same if the opponent sticks the pistol under the Marine's chin. To execute the counter when an opponent is pointing a pistol toward the front of a Marine, Marines—

Place the hands close to the weapon, about chest high, palms out.

Use the left hand to grab the opponent's forearm and push the opponent's hand with the pistol away to clear the body from in front of the weapon. At the same time, rotate the right shoulder back to clear the body from the weapon.

Maintain control of the opponent's arm.

Grasp the weapon with the right hand by placing the thumb underneath the pistol and the fingers over top of the pistol.

Keep the right hand wrapped tightly around the muzzle and quickly rotate the pistol in the opponent's hand so the muzzle is facing the opponent.

Grasp and pull the opponent's wrist or forearm away from the body while rotating the weapon.

Rotate the weapon toward the opponent while pulling it up and back and out of the opponent's grasp.

Pistol to the Rear

This technique is performed when Marines are unarmed and the opponent has a pistol pointing to the back of the Marine's head. To execute the counter when the opponent is pointing a pistol to the rear, Marines—

Place the hands close together about chest high, palms out.

Step back with the left foot, pivoting on the right foot so the side is against the opponent's front. This action clears the body from the weapon's line of fire. Keep the left hand up.

Pivot on the left foot to face the opponent and, at the same time, raise the left elbow and reach over the top of the opponent's arm with the left arm.

Wrap the left arm tightly around the opponent's arm above his elbow to control it.

Push on the opponent's shoulder with the right hand while pulling up with the left arm to achieve an armbar. This action releases the opponent's grip on the weapon. If necessary, the opponent can be taken to the ground with a leg sweep.

Note: To execute this technique, the weapon must be close to or touching the rear of the Marine. If the weapon is too far away from the body, this technique would be difficult to execute or it would be ineffective.

CHAPTER 3

HAND-HELD WEAPONS

This chapter describes all techniques for a right-handed person. However, all techniques can be executed from either side.

In drawings, the Marine is depicted in woodland camouflage utilities; the opponent is depicted without camouflage. In photographs, the Marine is depicted in woodland camouflage utilities; the opponent is depicted in desert camouflage utilities.

Marines must know how to defend against attacks when an opponent is either unarmed or armed with a held-held weapon. This chapter addresses the combative use of knives, specific weapons of opportunity, and sticks. However, virtually anything can be used as a hand-held weapon.

1. Fundamentals of Knife Fighting

Marines must be trained in knife fighting techniques. Marines experienced in offensive knife techniques can cause enough damage and massive trauma to stop an opponent. When engaged against each other, experienced knife fighters employ various maneuvers and techniques that are specific to knife fighting. Seldom, if ever, will Marines engage an opponent in a classical knife fight.

Note: When armed with a rifle, Marines are issued a bayonet. When armed with a pistol, Marine are issued a combat knife.

Angles of Attack

There are six angles from which an attack with a knife can be launched:

- Vertical strike coming straight down on an opponent.
- Forward diagonal strike coming in at a 45-degree angle to the opponent.
- Reverse diagonal strike coming in at a 45-degree angle to the opponent.
- Forward horizontal strike coming in parallel to the ground.
- Reverse horizontal strike coming in parallel to the ground.
- Forward thrust coming in a straight line to the opponent.

Target Areas of the Body

During any confrontation, the parts of the opponent's body that are exposed or readily accessible will vary. The goal in a knife fight is to attack the body's soft, vital target areas that are readily accessible (e.g, the face, the sides and front of the neck, the lower abdomen [or groin]).

Neck. Carotid arteries, located on either side of the neck, are good target areas because they are not covered by body armor or natural protection.

Lower Abdomen (or Groin). The lower abdomen (or groin region) is a good target area because it is not covered by body armor.

Heart. The heart, if not covered by body armor, is an excellent target which, if struck, can prove fatal in a matter of seconds or minutes.

Secondary Targets. There are secondary target areas that will cause substantial bleeding if an artery is severed. These target areas are not immediately fatal, but can become fatal if left unattended. Attacks to—

- The legs can cause a great deal of trauma and prove fatal. For example, the femoral artery located on the inside of the thigh is a large artery which, if cut, will cause extensive blood loss.
- The brachial artery, located between the biceps and triceps on the inside of the arm, can cause extensive bleeding and damage.
- The arm's radial and ulnar nerves can cause extensive bleeding and damage.

Movement

Marines can move anywhere within a 360-degree circle around the opponent. This allows accessibility to different target areas of the opponent's body. Marines should avoid being directly in front of an opponent because the opponent can rely on his forward momentum to seize the tactical advantage. If Marines face an opponent, movement is made in a 45-degree angle to either side of the opponent. This angle avoids an opponent's strike and places Marines in the best position to attack an opponent.

Wearing the Combat Knife

Marines must wear the combat knife where it is easily accessible and where it can best be retained. It is recommended the combat knife be worn on the weak side hip, blade down, sharp edge facing forward. Marines can place it behind the magazine pouch where it is easily accessible to them, but not easily grabbed by an opponent.

Grip

The grip on the knife should be natural. Marines grasp the knife's grip with the fingers wrapped around the grip naturally as it is pulled out of its sheath. This is commonly known as a hammer grip. The blade end of the knife is always facing the opponent.

Stance

Marines use the basic warrior stance as the foundation for knife techniques. The left hand is a vertical shield that protects either the ribs or the head and neck. The right elbow is bent with the blade pointing forward toward the opponent's head. This position serves as an index point, where all techniques are initiated.

Principles of Knife Fighting

The following are key principles of knife fighting:

Execute movements with the knife blade within a box, shoulder-width across from the neck down to the waistline. The opponent has a greater chance of blocking an attack if the blade is brought in a wide, sweeping movement to the opponent.
Close with the opponent, coming straight to the target.
Move with the knife in straight lines.
Point the knife's blade tip forward and toward the opponent.
Apply full body weight and power in each of the knife techniques. Full body weight should be put into the attack in the direction of the blade's movement (slash or thrust).
Apply constant forward pressure with the body and blade to keep the opponent off-balanced.

2. Knife Fighting Techniques

Slashing Techniques

Marines use slashing techniques to close with an enemy. Slashing techniques distract the opponent or damage the opponent so Marines can close in. Typically, Marines target the opponent's limbs, but any portion of the body that is presented can become a target.

Vertical Slash Technique. The vertical slash follows a vertical line straight down through the target. To execute the vertical slash, Marines—

Thrust the right hand out and bring the weapon straight down on the opponent, continuing to drag the knife down through the opponent's body.

Maintain contact on the opponent's body with the blade of the knife.

Forward Slash Technique. The forward slash follows a straight line in a forehand stroke, across the target areas of either the neck (high diagonal slash) or abdominal region (low horizontal slash). To execute the forward slash, Marines-
Extend the right hand while simultaneously rotating the palm up until the knife blade makes contact with the opponent.

Snap or rotate the wrist through the slashing motion to maximize blade's contact with the opponent.

Drag the knife across the opponent's body, from right to left, in a forehand stroke. The movement ends with the forearm against the body and the knife at the left hip with its blade oriented toward the opponent.

Reverse Slash Technique. The reverse slash is a follow-up technique to a forward attack. It allows Marines both a secondary attack and the ability to resume the basic warrior stance. The reverse slash follows a straight line in a backhand stroke, across the target areas of either the neck (high diagonal slash) or abdominal region (low horizontal slash). To execute the reverse slash, Marines—

Extend the right hand while simultaneously rotating the palm down until the knife blade makes contact with the opponent.

Snap or rotate the wrist through the slashing motion to maximize the blade's contact with the opponent.

Drag the knife across the opponent's body, from left to right, in a backhand stroke. Maintain contact on the opponent's body with the blade of the knife.

Thrusting Techniques

The primary objective of knife fighting is to insert the blade into an opponent to cause extensive damage and trauma. This is done with a thrusting technique. Thrusting techniques are more effective than slashing techniques because of the damage they can inflict. However, Marines use slashing techniques to close with the enemy so that they are closer to the opponent, which allows them to use the thrusting technique.

Vertical Thrust. The thrusting motion follows a vertical line straight up through the target (low into the abdomen region or high into the neck). To execute the vertical thrust, Marines—

Thrust the right hand toward the target, inserting the knife blade straight into the opponent.

Pull the knife out of the opponent.

Forward Thrust. The forward thrust follows a straight line straight into the opponent's neck (high thrust) or abdominal region (low thrust). To execute the forward thrust, Marines—

Thrust the right hand, palm down, toward the target, inserting the knife blade straight into the opponent.

Rotate the palm up once the knife is inserted to twist the blade.

Drop the right elbow and bring the knife to the opposite side of the opponent's body from where it was inserted. At the same time, rotate the hips and shoulders downward to bring body weight to bear on the attack.

Reverse Thrust. The reverse thrust is a follow-up technique to a forward attack. It allows Marines both a secondary attack and the ability to resume the basic warrior stance. The reverse thrust follows a horizontal line straight into the opponent's neck (high thrust) or abdominal region (low thrust). To execute the reverse thrust, Marines-

Bend the right arm, crossing the arm to the left side of the body.

Thrust the right hand, palm up, toward the target, inserting the knife blade straight into the opponent.

Rotate the palm down to twist the blade once the knife is inserted.

Bring the knife to the opposite side of the opponent's body from where it was inserted.

3. Weapons of Opportunity

During an unarmed close combat situation, Marines use their bodies as weapons, but they should be ready and able to use anything around them as a weapon. For example, Marines could throw sand or liquid in an opponent's eyes to temporarily impair his vision or smash the opponent's head with a rock or helmet. Marines must use whatever means are available and do whatever it takes to take control of the situation and to win, or they face the possibility of losing their lives. Some weapons of opportunity are discussed in the following subparagraphs.

Entrenching Tool

An entrenching tool (E-tool) is commonly carried by Marines. It can be an excellent weapon, especially when sharpened. Marines can use the E-tool to block, slash, and thrust at an opponent.

Tent Pole and Pins

Marines can use tent poles and pins to block, strike, or thrust at an opponent.

Web Belt

Marines can stretch a web belt between their hands to block attacks by an opponent.

Battlefield Debris

Marines can use debris on the battlefield (e.g., sticks, glass, a sharp piece of metal) to cut, slash, or stab an opponent. They can also use other types of debris such as shovel or ax handles, boards, metal pipes, or broken rifles to strike an opponent or apply a choke.

Helmet

A helmet can be used to strike an opponent on an unprotected area like the head and face. Grasp the rim of the helmet and thrust the arms forward, striking the opponent with the top of the helmet.

4. Fundamentals of Combative Stick

On the battlefield, Marines must be ready and able to use anything as a weapon. They must learn and be able to use techniques that can be employed with most weapons of opportunity. Among these techniques are combative stick techniques. Combative stick techniques can be used with a stick, a club, a broken rifle, an E-tool, or even a web belt.

Angles of Attack

There are six angles from which an attack with a hand-held weapon can be launched:

Vertical strike coming straight down on an opponent.
Forward diagonal strike coming in at a 45-degree angle to the opponent.
Reverse diagonal strike coming in at a 45-degree angle to the opponent.
Forward horizontal strike coming in parallel to the ground.
Reverse horizontal strike coming in parallel to the ground.
Forward thrust coming in a straight line to the opponent.

Grip

Grasp the stick about 2 inches from its base.

Stance

The basic warrior stance serves as the foundation for combative stick techniques. The left hand becomes a vertical shield that protects the ribs or the head and neck. Depending on how heavy the weapon is, it should be held at a level approximately shoulder height.

Movement

Movement during combative stick techniques is the same as it is for other close combat techniques. Marines can move anywhere within a 360-degree circle around the opponent. This allows accessibility to different target areas of the opponent's body and gains a tactical advantage. Marines should avoid being directly in front of an opponent because the opponent can rely on his forward momentum to seize the tactical advantage. If Marines face an opponent, movement is made in a 45-degree angle to either side of the opponent. This angle avoids an opponent's strike and places Marines in the best position to attack an opponent.

5. Combative Stick Techniques

Strikes

Strikes are intended to inflict as much damage on an opponent as possible. Striking techniques apply to a weapon of opportunity such as a stick, a tent pole, a club, a broken rifle, an E-tool, or a Pipe-

Vertical Strike. To execute the vertical strike, Marines—

Bend the right arm, extending the weapon over the back of the right shoulder.

Rotate the forearm straight down off the elbow to bring the weapon down on the opponent.

Rotate the hips and shoulders forcefully toward the opponent.

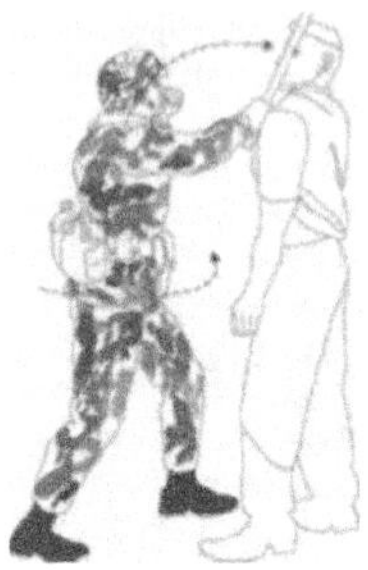

Follow through with the strike by allowing the weight of the weapon to go through the target area of the body.

Forward Strike. To execute a forward strike, Marines—

Step forward with the left foot in the direction of the strike.

Bend the right arm with the elbow extending out to the right and the weapon extended over the right shoulder.

Rotate the forearm to the right of the elbow to bring the weapon down onto the opponent. At the same time, forcefully rotate the hips and shoulders toward the opponent.

Follow through with the strike by allowing the weight of the weapon to go through the target area of the body.

Reverse Strike. The reverse strike is a follow-up technique to a forward strike. It allows Marines both a secondary attack and the ability to resume the basic warrior stance.To execute a reverse strike, Marines—

Step forward with the right foot in the direction of the strike.

Bend the right arm with the hand near the left shoulder. The weapon is extended over the left shoulder.

Rotate the forearm to the right of the elbow to bring the weapon down onto the opponent. At the same time, forcefully rotate the hips and shoulders toward the opponent.

Follow through with the strike by allowing the weight of the weapon to go through the target area of the body.

Forward Thrust. To execute the forward thrust, Marines—

Grasp the stick with the left hand, palm up, in a position where the stick can be controlled with two hands.

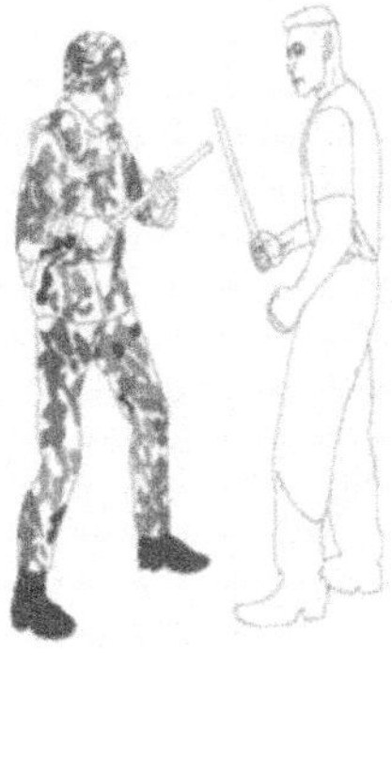

Lift the left leg and lunge forward off the ball of the right foot. At the same time, thrust the end of the weapon directly toward the opponent by thrusting both hands forward in a straight line.

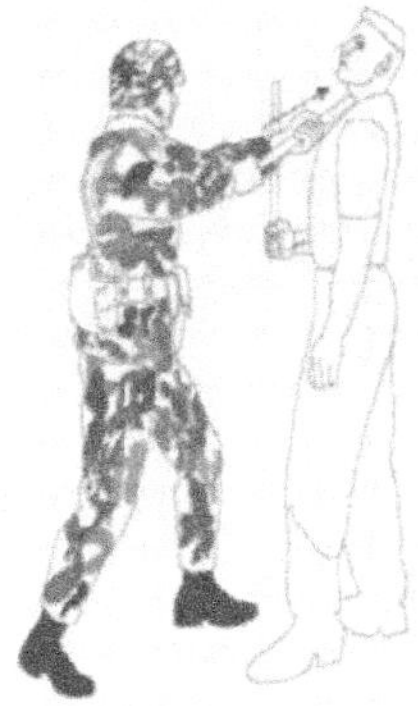

6. Blocking Techniques

A block is meant to deter or deflect an attack by an opponent. A block sets up Marines for a follow-on attack against the opponent. Blocks are executed by deflecting, rather than hitting or following through like a strike.

Blocks Against Unarmed Attacks

To block against an unarmed attack, Marines—

Step forward at a 45-degree angle with either the right or left foot. This moves the body out of the line of attack.

Raise the left arm and block the strike with the meaty portion of the forearm.

Employ the stick by one of two techniques:

Strike the opponent with the stick.
Use the stick to block on two points of contact. When the stick is used to block, it serves as an incidental strike. This technique is only used if Marines have closed with the opponent and are inside his strike.

Blocks Against Armed Attacks

Block for a Vertical Strike. To execute the block for a vertical strike, Marines—

Step forward with the left foot at a 45-degree angle to the left. This moves the body out of the line of attack.

Block on two points of contact to disperse the impact of the attack:

Block the opponent's stick by positioning the stick so it is perpendicular to the opponent' s stick. If the stick is not perpendicular to the opponent's stick, the stick can slide through and make contact on the Marine.
Block the opponent's wrist or forearm with the back of the left forearm.

Note: Use the stick to block the opponent's arm if closer to the opponent. It is the same movement as blocking with the arm, except the opponent's arm is blocked with both the stick and the arm.

Block for a Forward Strike. To execute the block for a forward strike, Marines-

Step forward with the left foot at a 45-degree angle to the left. This moves the body out of the line of attack and inside the opponent's strike.

Block on two points of contact to disperse the impact of the attack:

- Block the opponent's wrist or forearm with the meaty portion of the back of the left forearm.
- Strike the opponent's attacking biceps with the stick.

Block for a Reverse Strike. To execute the block for a reverse strike, Marines—

Step forward at a 45-degree angle with the right foot to the right. This moves the body out of the line of attack and inside the opponent's strike.

Block on two points of contact to disperse the impact of the attack:

Block the opponent's stick by positioning the stick so it is perpendicular to the opponent's stick.

Block the opponent's forearm with the meaty portion of the left forearm.

Note: If closer to the opponent, block his triceps with the back of the left forearm and strike his forearm with the stick.

7. Unarmed Against Hand-Held Weapons

If Marines are engaged against an opponent with a knife, a stick, or some other weapon of opportunity, they must establish and maintain an offensive mindset, not a defensive mindset. Their survival depends on it. They cannot afford to think about getting cut or hurt.

Angles of Attack

Before Marines can learn to block or counter an attack with a hand-held weapon (e.g., knife, stick), they must know from what angle the opponent is attacking. There are six angles from which an opponent will typically attack with a hand-held weapon:

Vertical strike coming in straight down on the Marine.
Forward diagonal strike coming in at a 45-degree angle to the Marine.
Reverse diagonal strike coming in at a 45-degree angle to the Marine.
Forward horizontal strike coming in parallel to the ground.
Reverse horizontal strike coming in parallel to the ground.
Forward thrust coming in a straight line to the Marine.

Blocks

If the opponent has a hand-held weapon (e.g., knife or stick), Marines parry the opponent's hand or arm to block the attack.

Basic Block. To execute the basic block technique, Marines-

Step forward at a 45-degree angle to move out of the line of the attack. Always step in the direction of the strike.

Thrust the forearms forward, hands up, against the opponent's attacking arm. Contact is made on the opponent's arm with the backs of the forearms.

Block for a Vertical Strike. To execute the block against a vertical strike, Marines—

Step forward with the left foot at a 45-degree angle to the left to move out of the line of attack.

Thrust the forearms forward, hands up, against the outside of the opponent's attacking arm.

Block for a Forward Strike. To execute the block against a forward diagonal or forward horizontal strike, Marines-

Step forward with the left fool inside the opponent's attacking arm.

Block the attack with both arms bent so the forearms make contact with the opponent's biceps and forearm.

Block for a Reverse Strike. To execute the block against a reverse diagonal or reverse horizontal strike, Marines—

Step forward with the right foot to the outside of the opponent's attacking arm.

Block the attack with both arms bent so the forearms make contact with the opponent's triceps and forearm.

Block for a Forward Thrust. To execute the block against a forward thrust, Marines—

Bend at the waist, move the hips backward, and jump backward with both feet to move away from the attack. This action is known as "hollowing out."

Hollow out and block the attack with the arms bent and hands together on top of the attacking arm.

Overlap the hands slightly so one thumb is on top of the other hand's index finger. The other thumb should be under the other hand's index finger.

8. Counters to Hand-Held Weapon Attacks

Principles of Counters

A counter is used to control the situation to regain the tactical advantage and end the fight. Regardless of the type of weapon or angle of attack, the following principles apply to countering the attack with a hand-held weapon:

Move out of the line of attack. Movement is executed in a 45-degree angle forward to the left or right.

- Block the attack.

Note: The first two actions are taken simultaneously.

- Take control of the weapon by controlling the hand or arm that is holding the weapon. Never attempt to grab the opponent's weapon.
- Execute the appropriate follow-up to end the fight; e.g., strikes, joint manipulations, throws, or takedowns (see chap. 8). Marines should continue to attack the opponent until the fight ends.

Counter Techniques

There are two techniques that can be used to counter any armed attack: forward armbar counter and reverse armbar counter. These techniques can be used to counter a vertical attack, a forward diagonal strike, or a forward horizontal strike. With minor variations, the same techniques are used to counter reverse strikes. A third technique, the bent armbar counter, is used to counter a vertical attack.

Forward Armbar Counter. To execute the forward armbar counter to an attack coming from a forward strike, Marines-

Step forward with the left foot inside the opponent' s attacking arm.

Block the attack with both arms bent so the forearms make contact with the opponent's biceps and forearm.

Slide the left arm over the opponent's forearm and wrap the arm tightly around his arm, trapping the opponent's attacking arm between the biceps and torso.

Place the right hand on the opponent's shoulder or upper arm to further control his arm and to effect an armbar.

Execute an armbar and continue to exert steady pressure against the arm.

Reverse Armbar Counter. To execute the reverse armbar counter to an attack coming from a forward strike, Marines-

Step forward with the left foot inside the opponent' s attacking arm.

Block the attack with both arms bent so the forearms make contact with the opponent's biceps and forearm.

Control the opponent's arm with the left hand and pivot to the right so the back is against the opponent's side. Immediately slide the right arm over the opponent's biceps and wrap the arm tightly around his arm, trapping the opponent's attacking arm between the biceps and torso.

Grasp the opponent's wrist with the left hand and twist his thumb away from the body.

Take the opponent to the ground with an armbar takedown if he does not drop the weapon from the wristlock.

Bent Armbar Counter. This counter is particularly effective against a vertical attack. To execute the bent armbar counter, Marines—

Step forward with the left foot inside the opponent's attacking arm.

Block the attack with both arms bent so the forearms make contact with the opponent's biceps and forearm.

Grasp the opponent's forearm with the left hand. At the same time, slide the right arm underneath the opponent's triceps and grasp the opponent's forearm or wrist with the right hand.

Apply downward pressure with the hands against the opponent's forearm to off-balance the opponent.

(reverse blank)

CHAPTER 4

STRIKES

This chapter describes all techniques for a right-lmnded person. However, all techniques can be executed from either side.

In drawings, the Marine is depicted in woodland camouflage utilities; the opponent is depicted without camouflage. In photographs, the Marine is depicted in woodland camouflage utilities; the opponent is depicted in desert camouflage utilities.

Strikes are unarmed individual hitting techniques. Strikes use the hands, elbows, knees, feet, and, in some instances, other parts of the body as personal weapons. Marines must know how to execute strikes effectively. They must also know how to counter strikes from an opponent.

1. Principles of Punches

Muscle Relaxation

Muscle relaxation is crucial when executing punches. The natural tendency in a fight is to tense up, which results in rapid fatigue and decreased power generation. Marines who remain relaxed during a close combat situation generate greater speed, which results in greater generation of power. Relaxing the forearms generates speed and improves reaction time. At the point of impact, Marines clench the fist to cause damage to the opponent and avoid injury to the wrist and hand.

Weight Transfer

Weight transfer is necessary to generate power in a punch. Marines accomplish this by-

Rotating their hips and shoulders into the attack.

Moving their body mass straight forward or backward in a straight line.

Dropping their body weight into an opponent. Body mass can be transferred into an attack from high to low or from low to high.

Rapid Retraction

When Marines deliver a punch, rapid retraction of the fist is important. Once the hand has made contact with the target, Marines quickly return to the basic warrior stance. Rapid retraction-

Returns the hand and arm to the protection afforded by the basic warrior stance.

Prevents the opponent from being able to grab the hand or arm.

Permits the hand and arm to be "chambered" or "re-cocked" in preparation for delivering a subsequent punch.

Telegraphing

Telegraphing a strike occurs when body movements inform the opponent of the intention to launch an attack. Staying relaxed helps to reduce telegraphing.

Often, an untrained fighter telegraphs his intention to attack by drawing his hand back in view of his opponent, changing facial expression, tensing neck muscles, or twitching. These movements, however small, immediately indicate an attack is about to be delivered. If the opponent is a trained fighter, he may be able to evade or counter the attack. If the opponent is an untrained fighter, he may still be able to minimize the effect of an attack.

2. Punches

Punches may be thrown during any hand-to-hand confrontation. Most people resort to punching because it is a natural reaction to a threat. The purpose of a punch is to stun the opponent or to set him up for a follow-up finishing technique. However, punches should only be executed to the soft tissue areas of an opponent. A correctly delivered punch maximizes the damage to an opponent while minimizing the risk of injury to Marines.

Basic Fist

Punches are executed using the basic fist. To make the basic fist, the fingers are curled naturally into the palm of the hand and the thumb is placed across the index and middle fingers. Do not clench the fist until movement has begun. This reduces muscular tension in the forearm and increases speed and reaction time. Just before impact, Marines exert muscular tension on the hand and forearm to maximize damage to the opponent and reduce their chances of injury. Contact should be made with the knuckles of the index and middle fingers.

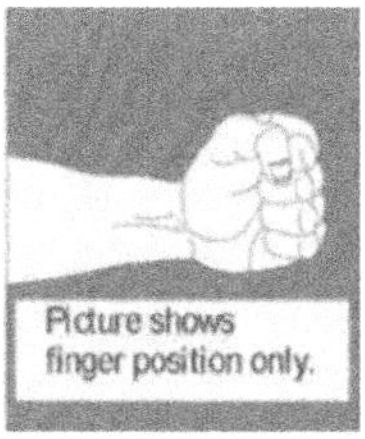

When striking with the basic fist, Marines must keep the hand straight, or in line, with the wrist to avoid injury to the wrist.

Lead Hand Punch

The lead hand punch is a snapping straight punch executed by the forward or lead hand. It is a fast punch designed to keep the opponent away and to set up an attack. A lead hand punch conceals movement and allows Marines to get close to the opponent. Lead hand punches should strike soft tissue areas, if possible. To execute the lead hand punch, Marines-

Snap the lead hand out to nearly full extension, while rotating the palm to the ground.

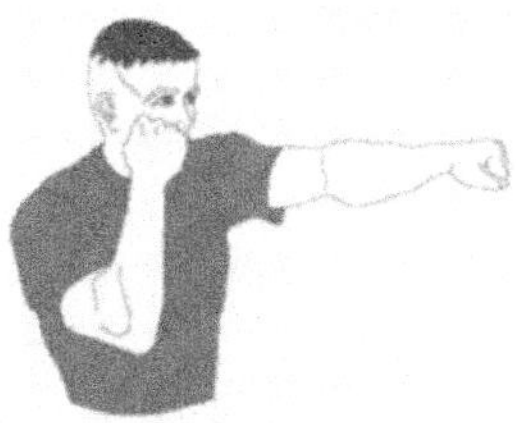

Contact the opponent with the first two knuckles of the fist.

Retract the hand immediately, resuming the basic warrior stance.

Rear Hand Punch

The rear hand punch is a snapping punch executed by the rear (right) hand. It is a power punch designed to inflict maximum damage on the opponent. Its power comes from pushing off the rear leg and rotating the hips and shoulders. To execute the rear hand punch, Marines—

Rotate the hips and shoulders forcefully toward the opponent and thrust the rear hand straight out, rotating the palm down, to nearly full extension.

Shift body weight to the lead foot while pushing off on the ball of the rear foot.

Contact the opponent with the first two knuckles of the fist.

Retract the hand immediately.

Uppercut

The uppercut is a powerful punch originating below the opponent's line of vision. It is executed in an upward motion traveling up the centerline of the opponent's body. It is delivered in close and usually follows a preparatory strike that leaves the target area unprotected. When delivered to the chin or jaw, the uppercut can render an opponent unconscious, cause extensive damage to the neck, or sever the tongue. To execute the uppercut, Marines—

Bend the arms, rotating the palm inboard. The distance the arms bend depends on how close the opponent is.

Rotate the hips and shoulders forcefully toward the opponent, thrusting the fist straight up toward the opponent's chin or jaw.

Contact the opponent with the first two knuckles of the fist.

Retract the hand immediately.

Hook

The hook is a powerful punch that is executed close in and is usually preceded by a preparatory strike. To execute the hook, Marines—

Thrust the right arm in a hooking motion toward the opponent, keeping the elbow bent while forcefully rotating the right shoulder and hip toward the opponent.

Contact the opponent with the first two knuckles of the fist. Continue rotating the shoulder and hip, following through with the fist to the target.

Retract the hand immediately.

3. Strikes with the Upper Body

Strikes stun the opponent or set him up for a follow-up finishing technique. The hands, forearms, and elbows are individual weapons of the arms that can be used to execute strikes including the hammer fist, chin jab, knife hand, eye gouge, and elbow strikes. These strikes provide a variety of techniques that can be used in any type of close combat situation.

Principles of Execution

There are several principles of execution that ensure a strike's effectiveness.

Generating Power. Weight transfer is necessary to generate maximum power in a punch. Marines accomplish this by-

Rotating the hips and shoulders into the attack.

Moving the body mass straight forward or backward in a straight line.

Dropping body weight into an opponent. Body mass can be transferred into an attack from high to low or from low to high.

Muscular Tension. The arms are relaxed until the moment of impact. At the point of impact, Marines apply muscular tension in the hand and forearm to maximize damage to the opponent and to avoid injury to the hand. The arms are relaxed until the moment of impact.

Hit and Stick and Follow-Through. A strike should be delivered so that the weapon (e.g., hand, elbow) hits and remains on the impact site (target) and follows through the target. This technique inflicts maximum damage on the opponent.

Strikes with the arms are executed with "heavy hands"; i.e., the strike is executed by driving through with the strike to allow the weight of the hand to go through the target area of the body. Contact with the opponent should be made with the arm slightly bent, and the arm extends as it moves through the target. This technique allows Marines to deliver strikes effectively without executing full force.

Movement. Movement puts Marines in the proper position for launching an attack against an opponent as well as providing protection.

Movement is initiated from the basic warrior stance and ends by resuming the basic warrior stance. Strikes can be performed with either the left or right arm depending upon—

The angle of attack.
The position of the opponent.
The opponent's available, vulnerable target areas.

Target Areas of the Body

For each strike, there are target areas of the body that, when struck, maximize damage to an opponent. Strikes use gross motor skills as opposed to fine motor skills. The target areas of the body are just that—areas. Pinpoint accuracy on a specific nerve is not needed for the strike to be effective.

Hammer Fist

Striking with the hammer fist concentrates power in a small part of the hand, which, when transferred to the target, can have a devastating effect. The striking surface of the hammer fist is the meaty portion of the hand below the little finger. To execute the hammer fist strike, Marines-

Make a fist and bend the arm at approximately a 45- to 90-degree angle. At the same time, rotate the right hip and right shoulder backward.

Thrust the fist forward onto the opponent while rotating the right hip and shoulder forward.

Rotate the wrist so the hammer fist makes contact on the opponent.

Chin Jab

The chin jab can immediately render an opponent unconscious and cause extensive damage to the neck and spine. The striking surface is the heel of the palm of the hand. To execute the chin jab, Marines—

Bend the right wrist back at a 90-degree angle with the palm facing the opponent and the fingers pointing up.

Keep the right arm bent and close to the body. Extend the hand into a concave position with the fingers slightly spread apart.

Step forward with the left foot toward the opponent, keeping the feet approximately shoulder-width apart and the knees bent. This is done to close with the opponent.

Keep the right arm bent and close to the side. Thrust the palm of the hand directly up under the opponent's chin. At the same time, rotate the right hip forward to drive the body weight into the attack to increase the power of the strike. The attack should travel up the centerline of the opponent's chest to his chin.

Knife Hand

The knife hand is one of the most versatile and devastating strikes. The striking surface is the cutting edge of the hand, which is the meaty portion of the hand below the little finger extending to the top of the wrist. The striking surface is narrow, allowing strikes on the neck between the opponent's body armor and helmet. The knife hand strike is executed from one of three angles: outside, inside, and vertical.

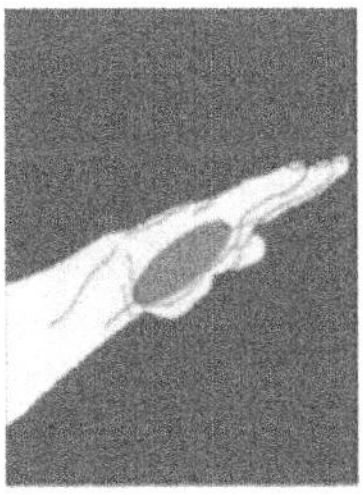

Outside Knife Hand. To execute the outside knife hand strike, Marines—

Execute a knife hand by extending and joining the fingers of the right hand and placing the thumb next to the forefinger (like saluting).

Retract the right hand. At the same time, rotate the right hip and right shoulder backward.

Thrust the knife hand forward (horizontally) onto the opponent while rotating the right hip and shoulder forward.

Inside Knife Hand. To execute the inside knife hand strike, Marines—

Execute a knife hand.

Bring the right hand over the left shoulder. At the same time, rotate the right shoulder forward and the left hip forward.

Thrust the knife hand forward (horizontally) onto the opponent while rotating the right hip and shoulder forward and the left shoulder backward.

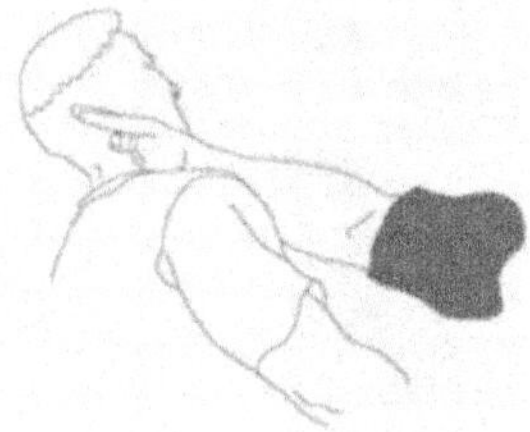

Vertical Knife Hand. When thrown vertically, the knife hand strike comes straight down in a straight line.

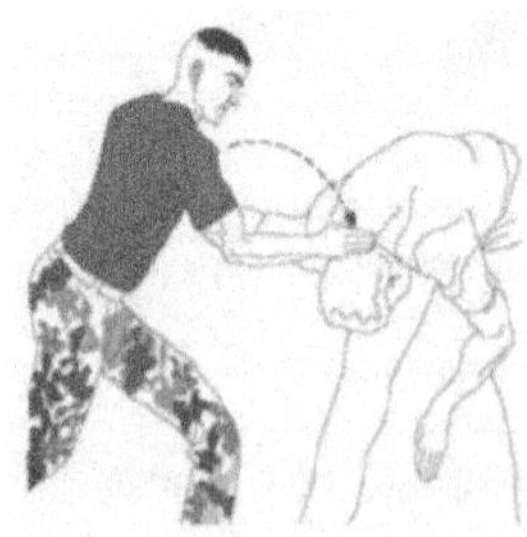

Eye Gouge

The eye gouge is used to attack an opponent's eyes, blinding him so follow-up strikes can be executed. The striking surface is the tips of the fingers or thumb. To execute the eye gouge, Marines—

Extend the right hand with the fingers slightly spread apart to allow entry into the eye sockets.

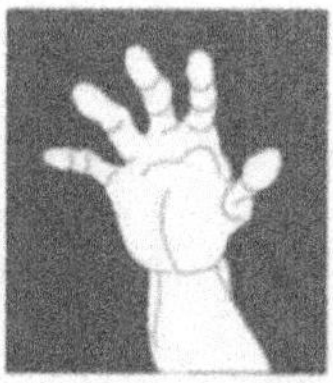

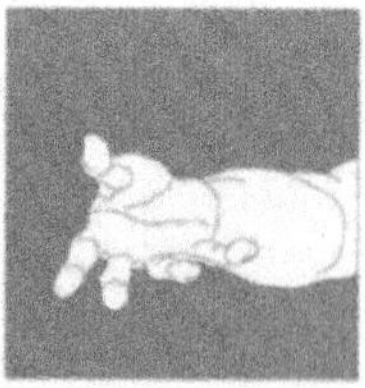

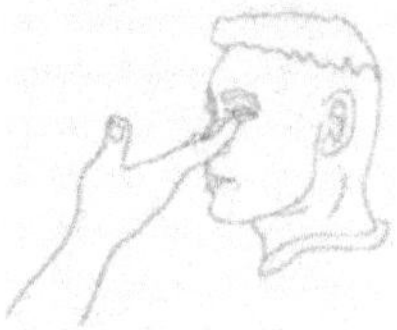

Place the palm of the hand either toward the ground or toward the sky and thrust the right hand forward into the opponent's eyes.

Thrust the hand forward at the opponent's nose level so the fingers or thumb slide naturally into the grooves of the opponent's eye sockets.

Elbow Strikes

The elbow is a powerful weapon that can be used in several different ways to attack virtually any part of an opponent's body. Elbow strikes can be performed either vertically (upward or downward) or horizontally (forward or reverse). The striking surface is 2 inches above or below the point of the elbow, depending upon the angle of attack, the opponent's attack angle, and the position of the opponent.

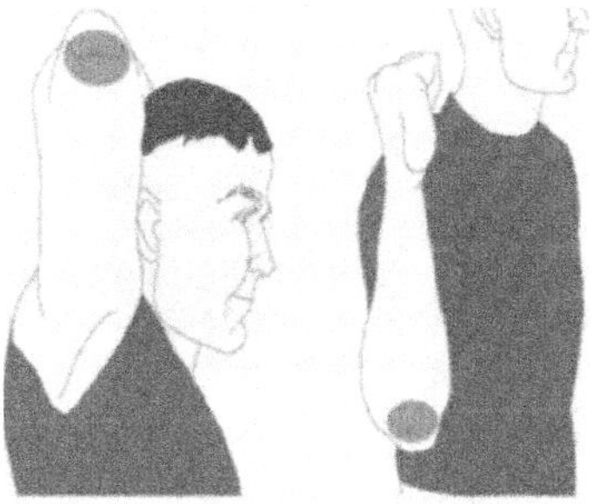

Above Elbow Below Elbow

Vertical Elbow Strike (Up). To execute an upward vertical elbow strike, Marines-

Bend the right elbow, keeping the fist close to the body. The fist is at shoulder level and the elbow is next to the torso.

Thrust the elbow vertically upward toward the opponent while rotating the right shoulder and hip forward to generate additional power.

Contact the opponent with the right forearm 2 inches above the point of the elbow.

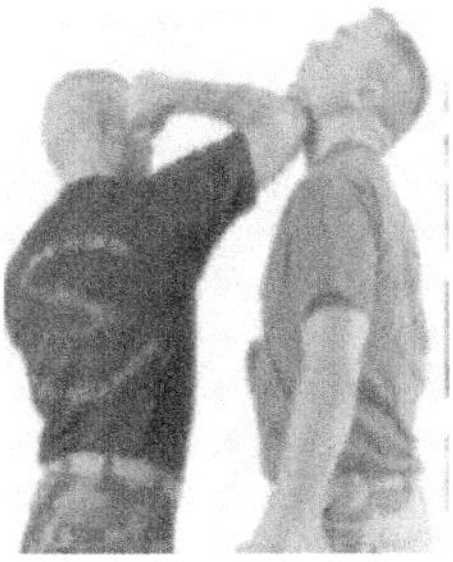

Vertical Elbow Strike (Down). To execute a downward vertical elbow strike, Marines—

Bend the right elbow, keeping the fist close to the body. The fist is on the shoulder and the elbow is raised well above the shoulder.

Thrust the elbow vertically downward toward the opponent while dropping body weight into the attack to generate additional power.

Contact the opponent with the right triceps 2 inches above the point of the elbow.

Horizontal Elbow Strike (Forward). To execute the forward horizontal elbow strike, Marines-

Tuck the right fist near the chest with the palm heel facing the ground.

Thrust the right elbow horizontally forward toward the opponent. The forearm is parallel to the ground.

Rotate the right shoulder and hip forward to generate additional power.

Contact the opponent with the right forearm 2 inches below the point of the elbow.

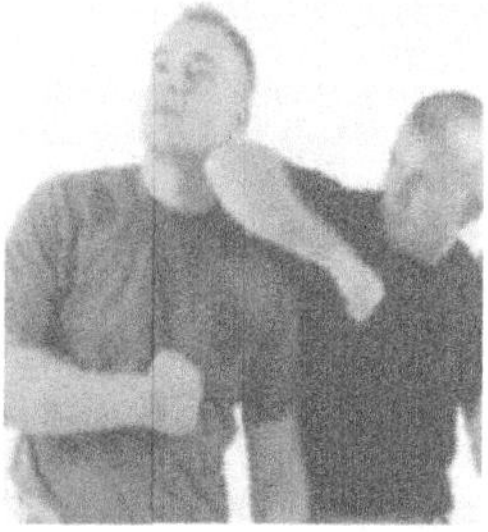

Horizontal Elbow Strike (Rear). To execute the rear horizontal elbow strike, Marines-

Tuck the right fist near the left shoulder with the palm heel facing the ground. At the same time, rotate the right shoulder forward and the left hip forward.

Thrust the right elbow horizontally rearward toward the opponent. The forearm is parallel to the ground and the hand moves toward the direction of the attack.

Rotate the right hip back and the right shoulder backward to generate additional power.

Contact the opponent with the right triceps 2 inches above the point of the elbow.

4. Strikes with the Lower Body

The legs are the body's most powerful weapons because they use the largest muscle groups to generate a strike. Legs are also less prone to injury. The feet are the preferred choice for striking because they are protected by boots. Marines use their feet, heels, and knees to execute kicks, knee strikes, and stomps.

Kicks

The purpose of a kick is to stop an opponent's attack or to create an opening in his defense in order to launch an attack. Kicks can be performed with the left (lead) leg or the right (rear) leg. Kicks with the rear leg have greater power because the hips are rotated into the attack. However, the rear leg is further away from the opponent so a strike with the rear leg will not contact the opponent as quickly as a strike with the lead leg.

Front Kick. The front kick is executed when the opponent is in front of the Marine. The front kick, delivered with either the rear or lead leg, is effective for striking below the waist. Attempting to kick higher results in diminished balance and provides the enemy with a greater opportunity to grab the leg or foot. The striking surfaces arc the toe of the boot or the bootlaces. To execute the front kick, Marines-

Raise the left knee waist high, pivot the hips into the attack, and thrust the left foot forward toward the opponent.

Contact the opponent with the toe of the left boot or bootlaces.

Return to the basic warrior stance.

Side Kick. The side kick, delivered with the lead leg, is effective for striking the knees. The side kick is executed when the opponent is to the side of the Marine. The striking surface is the outside cutting edge of the boot near the heel. To execute the side kick, Marines—

Raise the right knee waist high and rotate the right hip forward.

Thrust the right foot to the right side toward the opponent, turning the foot at a 90-degree angle to maximize the striking surface on the opponent.

Contact the opponent with the cutting edge of the right boot.

Return to the basic warrior stance.

Knee Strikes

Knee strikes are excellent weapons during the close range of close combat fighting. The knee strike is generally delivered in close.

Vertical Knee Strike. The striking surface is the thigh, slightly above the knee. To execute the vertical knee strike, Marines-

Raise the right knee and drive it up forcefully into the opponent. Power is generated by thrusting the leg upward.

Contact the opponent 2 inches above the right knee.

Horizontal Knee Strike. The horizontal knee strike is executed with the leg generally parallel with the ground while rotating the hips to generate power. It is often delivered to the peroneal nerve. The striking surface is the front of the leg, slightly above or below the knee. To execute the horizontal knee strike, Marines-

Raise the right knee, rotate the right hip forward while pivoting on the left foot, and drive the knee horizontally into the opponent.

Contact the opponent 2 inches above the right knee.

Stomps

Stomps are delivered with the feet, usually when the opponent is down. Remember, when the opponent is down, Marines take whatever target is available.

Vertical Stomp. The vertical stomp allows Marines to remain upright and balanced, to rapidly deliver multiple blows with either foot, and to quickly and accurately attack the target. It is the preferred stomp. The striking surface is the flat bottom of the boot or the cutting edge of the heel. To execute the vertical stomp, Marines—

Raise the right knee above the waist with the right leg bent at approximately a 90-degree angle.

Drive the flat bottom of the right boot or the cutting edge of the right heel down onto the opponent forcefully. At the same time, bend the left knee slightly to drop the body weight into the strike.

Ax Stomp. The striking surface of the ax stomp is the cutting edge of the heel. To execute the ax stomp, Marines-

Raise the right heel above the waist, keeping the right leg straight.

Drive the cutting edge of the right heel down onto the opponent forcefully. At the same time, bend the left knee slightly to drop the body weight into the strike.

5. Counters to Strikes

In a close combat situation, an opponent will attempt to strike Marines with punches and kicks. When an opponent uses a strike, Marines must first avoid the strike. This is accomplished by moving quickly and blocking. Next, Marines must get into an offensive position. This allows Marines to use offensive strikes to attack the opponent. Regardless of the strike, the counter to a strike requires Marines to move, block, and strike.

Move

The first step in countering a strike is to move out of the way of the strike's impact. Movement removes Marines from the opponent's intended strike point and positions Marines to attack. Movement is executed at approximately a 45-degree angle to the front or rear. Movement is always initiated from the basic warrior stance. Return to the basic warrior stance with the toe of the lead foot pointing toward the opponent once the movement is complete.

Block

Different blocks are executed based on the strike. These will be covered by individual counters.

Strike

Any of the upper body or lower body strikes can be executed as a follow-on attack or part of the counter to an opponent's strike. The follow-on strike is determined by the angle to the opponent, the position of the opponent, and the opponent's available vulnerable target areas.

Counters to Punches

Counter to a Lead Hand Punch. This counter is used when the opponent throws a lead hand punch. To execute the counter to the lead hand punch, Marines-

Step forward and to the left at approximately a 45-degree angle, moving in to the outside of the opponent's attacking arm.

Raise the left arm and block or deflect the opponent's lead hand with the palm of the hand or the meaty portion of the forearm.

Hit and stick by leaving the left arm against the opponent's right arm while stepping forward and to the left at approximately a 45-degree angle to close with the opponent.

Execute a finishing technique, such as a strike or a kick, to the opponent's exposed target areas.

Counter to a Rear Hand Punch. This counter is used when the opponent throws a rear hand punch. To execute the counter to the rear hand punch, Marines—

Step forward and to the left at approximately a 45-degree angle, moving in to the outside of the opponent's attacking arm.

Raise the left arm and block or deflect the opponent' s rear hand with the palm of the hand or the meaty portion of the forearm.

Hit and stick by leaving the left arm against the opponent's right arm while stepping forward and to the right at approximately a 45-degree angle to close with the opponent.

Execute a finishing technique, such as a strike or a kick, to the opponent's exposed target areas.

Counters to Kicks

Counter to a Front Kick (Left or Lead Leg). This counter is used when the opponent executes a front kick with his left leg. To execute the counter to a front kick, Marines-

Step forward and to the right at approximately a 45-degree angle, moving in to the outside of the opponent's striking leg.

Lower the left arm and block or deflect the opponent' s leg with the palm of the hand or the meaty portion of the forearm.

Hit and stick by leaving the left arm against the opponent's leg while stepping forward and to the left at approximately a 45-degree angle to close with the opponent.

Execute a finishing technique, such as a strike or a kick, to the opponent's exposed target areas.

Counter to a Front Kick (Right or Rear Leg). This counter is used when the opponent executes a front kick with his right leg. To execute the counter to a front kick, Marines—

Step forward and to the left at approximately a 45-degree angle, moving in to the outside of the opponent's striking leg.

Lower the left arm and block or deflect the opponent' s leg with the palm of the hand or the meaty portion of the forearm.

Hit and stick by leaving the left arm against the opponent's leg while stepping forward and to the right at approximately a 45-degree angle to close with the opponent.

CHAPTER 5

THROWS

This chapter describes all techniques for a right-handed person. However, all techniques can be executed from either side.

The Marine is depicted in camouflage utilities. The opponent is depicted without camouflage.

Marines use throwing techniques to maintain the tactical advantage and to throw the opponent to the ground during close combat. Throws apply the principles of balance, leverage, timing, and body position to upset an opponent's balance and to gain control by forcing the opponent to the ground. Throwing techniques are effective because they are size- and gender-neutral, and they rely on the momentum and power generated by the opponent rather than the strength or size of the Marine. Marines also execute a throw as a devastating attack against an opponent, possibly causing unconsciousness or broken limbs. When Marines execute throws, they must maintain balance and, simultaneously, prevent the opponent from countering a throw or escaping after being forced to the ground.

1. Turning Throw

Marines use a turning throw to take the opponent to the ground while they remain standing. A turning throw can also be executed from a stationary position. It is particularly effective if Marines and the opponent are wearing gear. To execute the turning throw, Marines-

Grasp the opponent's right wrist with the left hand.

Step forward with the right foot, place it against the outside of the opponent's right foot, and pivot so the back of the heel is next to the middle of the opponent's foot.

Hook the opponent's right arm with the right arm and pinch his arm between the biceps and forearm, touching the opponent with the body.

Pull the opponent's wrist downward, keeping it close to the body.

Pivot to the left on the ball of the foot and continue pulling downward on the opponent's wrist while rotating the wrist outward to off-balance him.

2. Hip Throw

Marines use a hip throw to take the opponent to the ground while they remain standing. A hip throw is particularly effective if the opponent is moving forward or pushing. Marines use the opponent's forward momentum to execute the hip throw. To execute the hip throw, Marines-

Grasp the opponent's right wrist with the left hand.

Step forward with the right foot and place it against the outside of the opponent's right foot.

The back of the heel should be next to the opponent's foot.

Step back with the left foot and rotate on the ball of the foot. The back of the heel is next to the opponent' s toe. The knees are bent.

Rotate at the waist and hook the right hand around the back of the opponent's body (anywhere from his waist to his head). The side and hip should be against the opponent.

Rotate the hip against the opponent. The hips must be lower than the opponent's.

Use the right hand to pull the opponent up on the hip to maximize contact.

Pull the opponent's arm across the body and, at the same time, lift the opponent off the ground slightly by bending at the waist, straightening the legs, and rotating the body to the left.

3. Leg Sweep

Marines use a leg sweep to take the opponent to the ground while they remain standing. A leg sweep is particularly effective if the opponent is already off-balanced and moving backward or pulling on the Marine. To execute the leg sweep, Marines—

Grab the opponent's right wrist with the left hand and grab the opponent' s left shoulder with the right hand.

Note: Grab the opponent's clothing or gear if his wrist and shoulder cannot be grabbed.

Step forward with the left foot and place it on the outside of the opponent's right foot.

Pull the opponent's wrist downward, close to the body, and push his shoulder backward to off-balance him.

Raise the right knee no higher than the waist.

Kick the foot past the opponent's right leg.

Use the heel of the boot to make contact with the opponent's calf (anywhere from the top of the calf down to the Achilles tendon or on the inside of the calf).

Sweep the opponent to the ground.

CHAPTER 6

CHOKES AND HOLDS

This chapter describes all techniques for a right-handed person. However, all techniques can be executed from either side.

The Marine is depicted in camouflage utilities. The opponent is depicted without camouflage.

WARNING

During training, never execute a choke at full force or full speed and never hold a choke for more than 5 seconds.

When Marines correctly perform a choke, they render an opponent unconscious in as little as 8 to 13 seconds. Chokes are easily performed regardless of size or gender. Marines must know how to apply chokes and how to counter a choke or a hold executed by an opponent.

1. Types of Chokes

There are two types of chokes: an air choke and a blood choke. An air choke closes off the airway to the lungs, thereby preventing oxygen from reaching the heart. A blood choke cuts off the blood flow to the brain. Both types can result in unconsciousness and eventual death for an opponent.

Air Choke

An air choke is performed on the opponent's windpipe (or trachea), cutting off the air to the lungs and heart. If Marines execute the air choke properly, the opponent loses consciousness within 2 to 3 minutes. Due to the length of time it takes to immobilize the opponent, air chokes are not recommended.

Blood Choke

A blood choke is performed on the opponent's carotid artery, which carries oxygen-enriched blood from the heart to the brain. The carotid artery is located on both sides of the neck. If Marines execute a blood choke properly, the opponent will lose consciousness within 8 to 13 seconds. The blood choke is the preferred choke because its intended effect (i.e., the opponent losing consciousness) can be executed quickly, ending the fight.

2. Chokes

Front Choke

Marines execute a front choke when they are facing the opponent. Marines use the opponent's lapels or collar to execute a front choke. To execute the front choke, Marines-

Grab the opponent's right lapel with the right hand, making certain that the knuckles or the back of the hand are against the opponent's carotid artery on the right side of his neck.

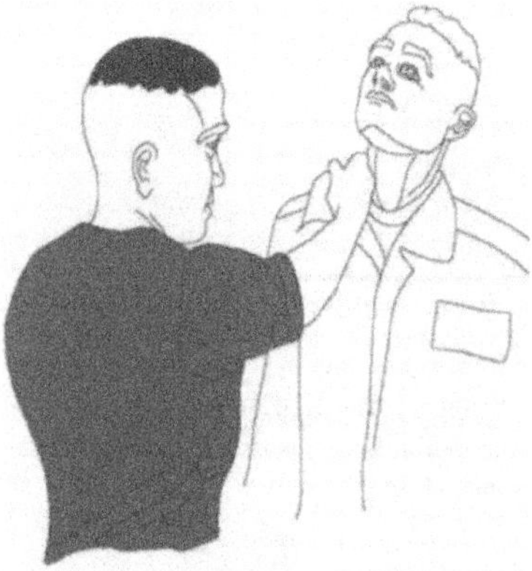

Keep the right hand against the opponent's neck, reach under the opponent's right arm with the left hand, grab the opponent's left lapel, and form an X with the wrists.

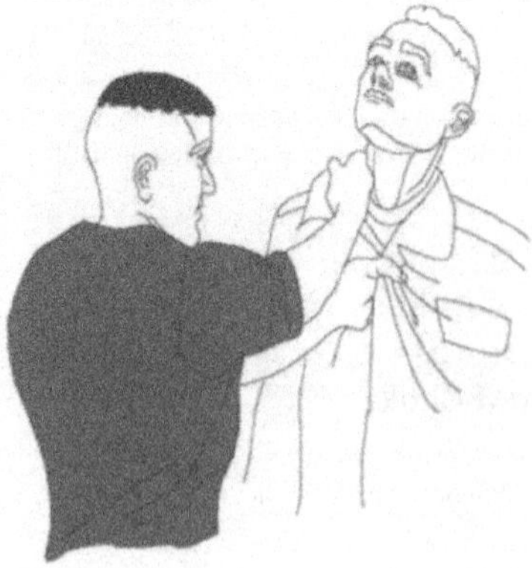

Press the right hand against the opponent's carotid artery.

Pull his lapel to the left with the left hand.

Side Choke

Marines execute a side choke when they are facing the opponent. The side choke is particularly effective when deflecting a punch thrown by an opponent. To execute the side choke, Marines—

Use the left hand to parry the opponent's arm inboard (to the inside of the opponent's reach).

Bring the right arm underneath the opponent's arm and up around the front of his neck.

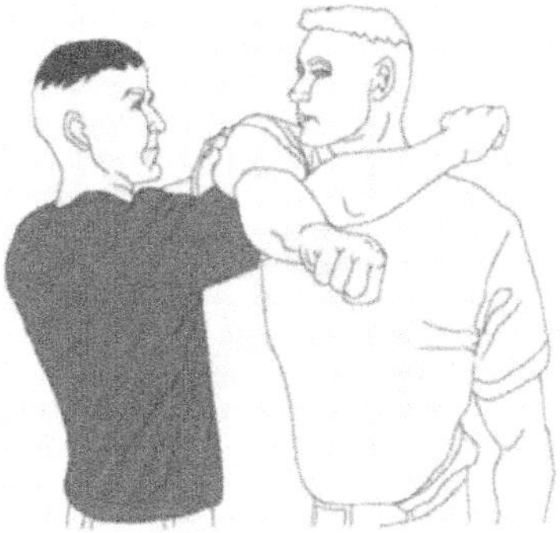

Extend the fingers, place the back of the forearm against the opponent's neck just below his ear, and press the carotid artery on his neck.

Note: If Marines are unable to place their fingers against the opponent's neck, they may use the back of the thumb or wrist.

Reach, with the left hand, around the back of the opponent's neck and clasp the hands together.

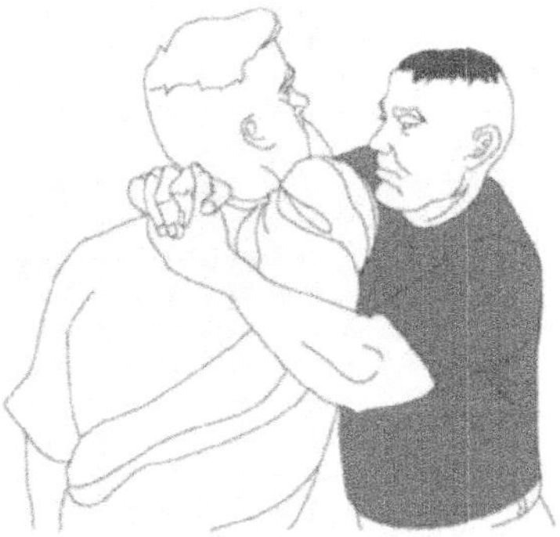

Pull the opponent toward the chest by pulling the clasped hands toward the chest.

Exert pressure on the side of the opponent's neck with the forearm.

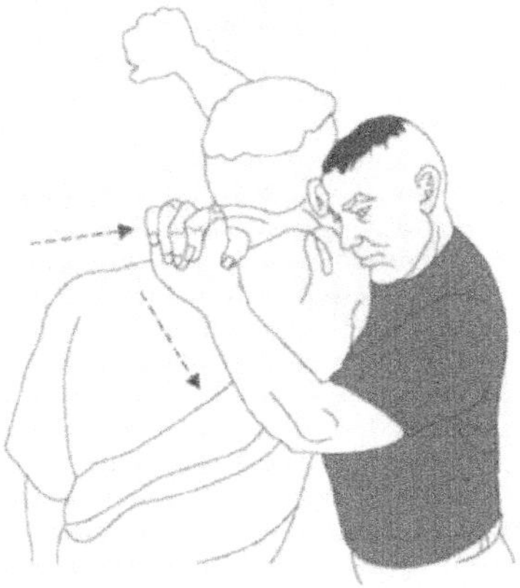

Rear Choke

Marines execute a rear choke when they are behind the opponent, the opponent is on the ground, or when they are taking the opponent to the ground. To execute the rear choke, Marines—

Reach, with the right arm, over the opponent's right shoulder and hook the bend of the arm around his neck.

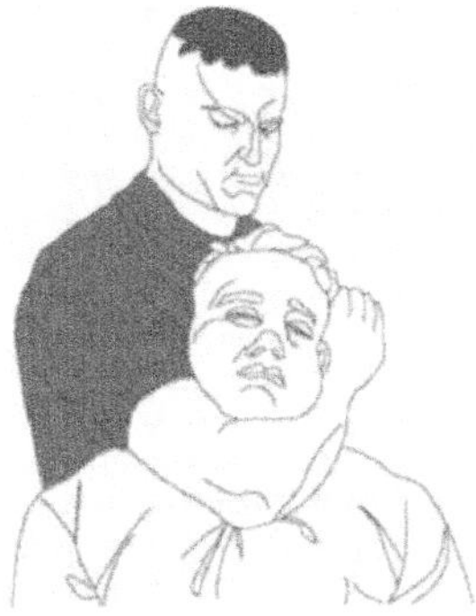

Clasp both hands together.

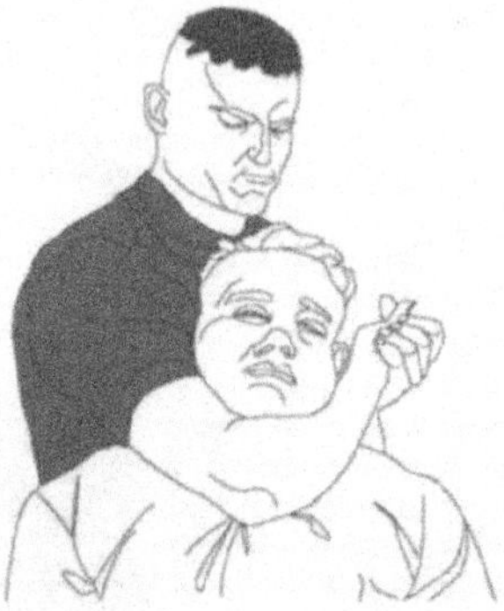

Exert pressure with the biceps and forearms on both sides of the opponent's neck on his carotid arteries.

Maintain pressure with the biceps and forearms on both sides of the neck and draw the opponent closer by drawing the right arm in.

Figure-Four Choke

A variation of the rear choke is the figure-four choke. The figure-four choke allows Marines to gain more leverage than the rear choke. If the rear choke cannot be secured, the figure-four variation is employed to increase the pressure of the choke on the opponent. To execute the figure-four variation of the rear choke, Marines—

Apply a rear choke. The Marine's body should be against the opponent's body.

Grasp the left biceps with the right hand and place the left hand against the back of the opponent's head.

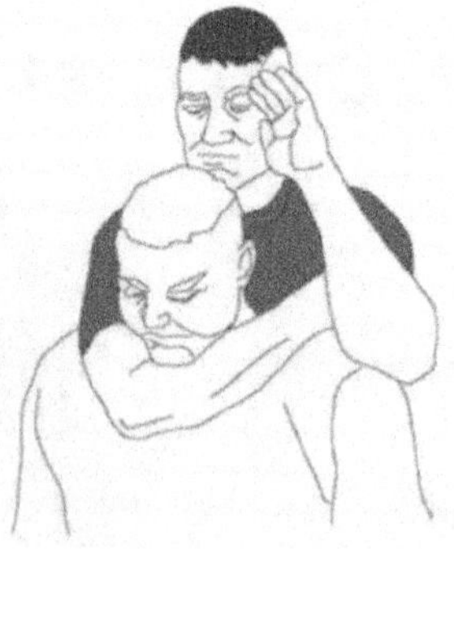

Push the opponent's head forward and down with the left hand.

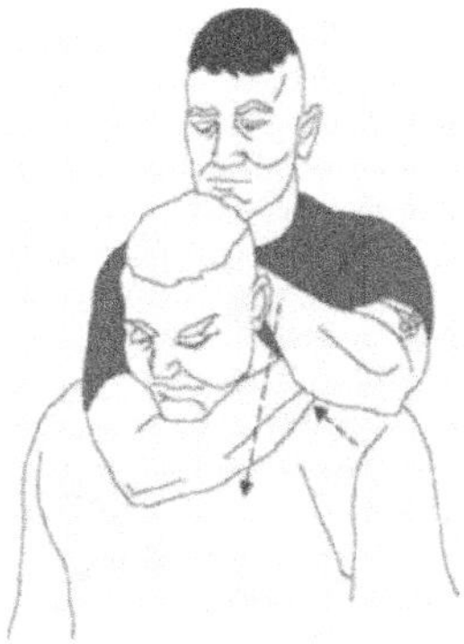

Draw the right arm in, maintaining pressure with the bicep and forearm on both sides of the opponent's neck.

3. Counters to Chokes and Holds

During a close combat situation, an opponent may apply a choke or hold on a Marine. If the opponent correctly applies a choke, a Marine quickly loses consciousness. If a choke is not executed properly, it often results in a hold, typically a bear hug or a headlock. A hold allows the opponent to control a Marine and removes the Marine's ability to attack. It is important for Marines to extract themselves from chokes and holds, regain the tactical advantage, and counter with strikes.

Although a choke causes unconsciousness in 8 to 13 seconds for a blood choke and 2 to 3 minutes for an air choke, the first movement in any counter to a choke is to clear the airway. Marines use softening techniques to loosen an opponent's grip and to clear their airway. Softening techniques are particularly effective if Marines lack the physical strength of their opponent. These techniques include groin strikes, eye gouges, foot stomps, etc. Softening techniques are not offensive; rather, they are used to loosen an opponent's hold.

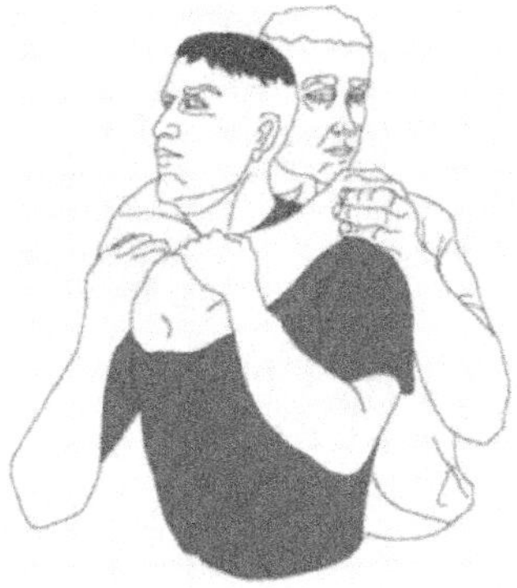

The second movement is to tuck the chin. Once the airway is clear, Marines tuck their chins to prevent the opponent from reapplying the choke.

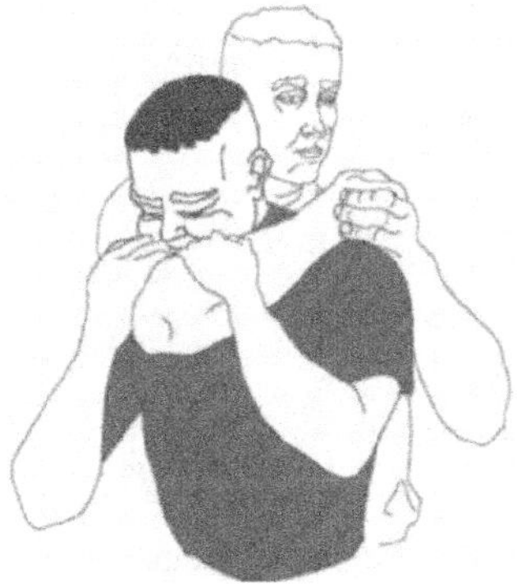

Counter to a Front Choke

Marines use a counter to a front choke when the opponent approaches from the front and uses both hands to choke a Marine around the throat. To execute the counter to the front choke, Marines—

Grasp the opponent's right forearm (where the elbow bends) with the left hand and apply downward pressure on the opponent's radial nerve with the fingers.

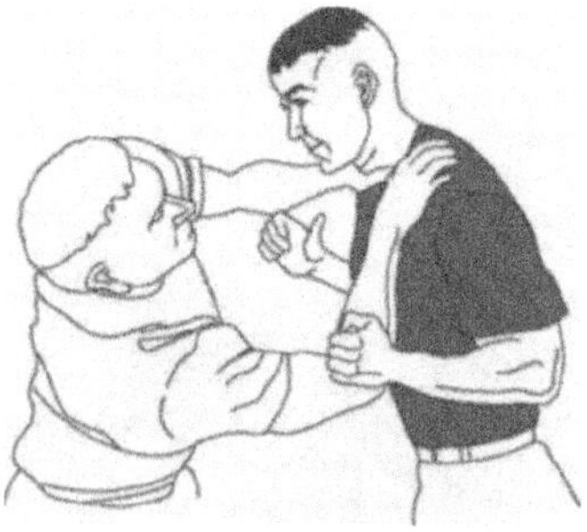

Execute a chin jab to the opponent's chin with the right hand. To generate power into the strike, bring the left foot to the outside of the opponent's right foot and rotate the hips into the strike.

Counter to a Rear Choke

Marines execute a counter to a rear choke when the opponent approaches from the rear and puts his right arm around a Marine's throat. To execute the counter to the rear choke, Marines-

Grasp the opponent's forearm (at the radial nerve) and bicep with both hands and pull down just enough to clear the airway. Once the airway is clear, tuck the chin to protect the airway and to prevent the opponent from reapplying the choke.

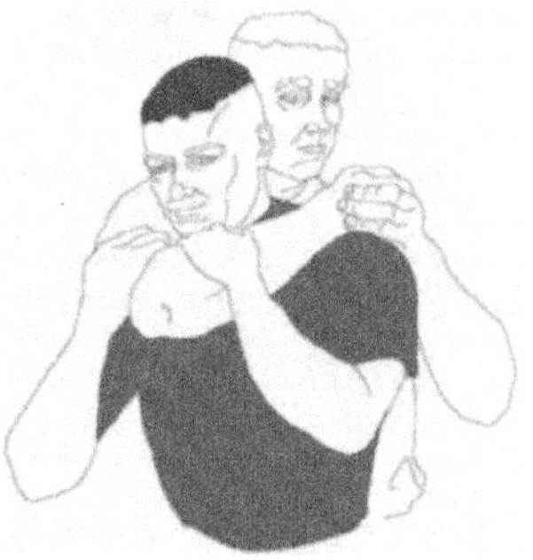

Step behind the opponent's right leg with the left foot, keeping both legs bent (almost in a squatting position).

Strike and drive the left elbow into the opponent's torso while rotating the hips and pivoting to the left, throwing the opponent back and to the ground.

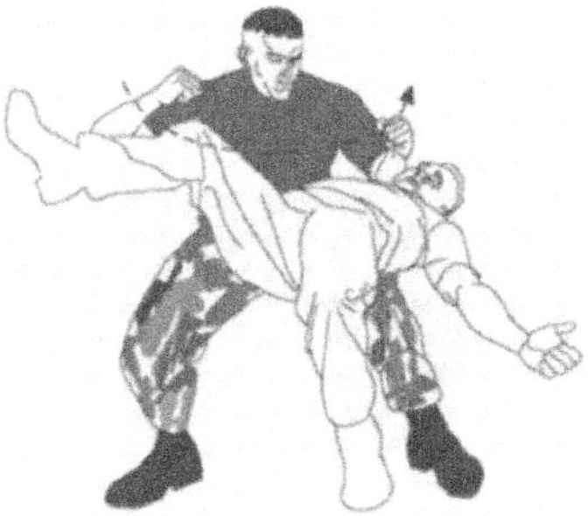

Counter to a Front Headlock

Marines use a counter to a front headlock when the opponent approaches from the front and puts his right arm around the Marine's neck, bends the Marine forward, and locks the Marine's head against his hip. To execute the counter to a front headlock, Marines—

Grasp the opponent's wrist and forearm with both hands and pull down to clear the airway. Maintain control of the opponent's wrist throughout the move. Once the airway is clear, tuck the chin to protect the airway and to prevent the opponent from reapplying the choke.

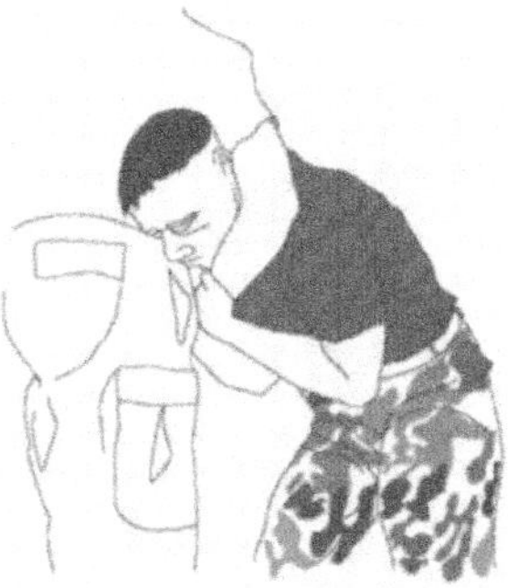

Move the right hand and arm across the opponent's torso.

Step forward and to the left with the left foot at a 45-degree angle.

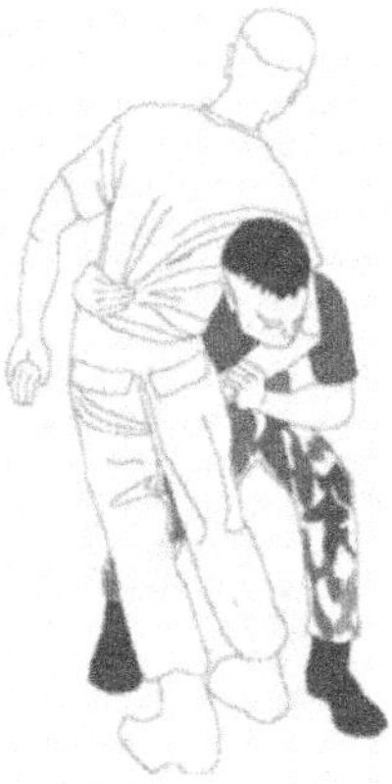

Execute a sweep with the right foot against the opponent's right leg. At the same time, push against the opponent's chest with the right arm and shoulder to generate power in the sweep.

Counter to a Rear Headlock

Marines use a counter to a rear headlock when the opponent approaches from the rear and puts his right arm around the Marine's neck, bends the Marine forward, and locks the Marine's head against his hip. To execute the counter to the rear headlock, Marines-
Grasp the opponent's wrist and forearm with the right hand and pull down to clear the airway.
Once the airway is clear, tuck the chin to protect the airway and to prevent the opponent from re-applying the choke.

Reach over the opponent's right shoulder with the left arm.

Grab any part of the opponent's face (chin, nose, eyes) and pull back while rising to a standing position.

Execute, with the right hand, a hammer fist strike to the opponent's exposed throat.

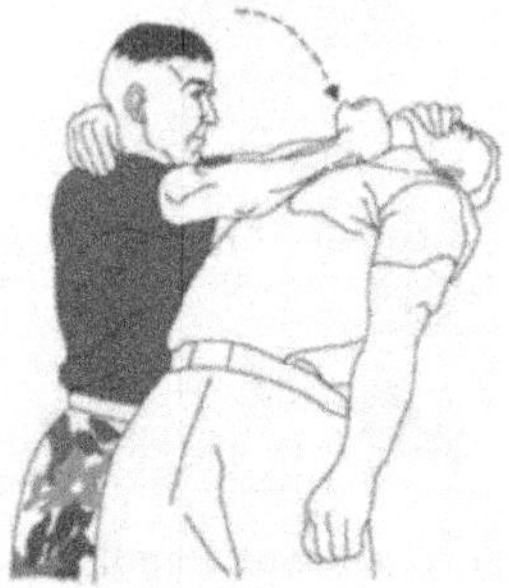

Counter to a Front Bear Hug

Marines execute a counter to a front bear hug when the opponent approaches from the front and puts both of his arms around the Marine's body, trapping the Marine's arms to the sides. To execute the counter to a front bear hug, Marines—

Step forward and to the left with the left foot at a 45-degree angle to the outside of the opponent's right leg, keeping the left leg bent.

Grasp the opponent's torso or arms to gain balance and to assist in throwing the opponent. It may be helpful to hook the opponent's right arm with the left arm.

Drive the right arm and shoulder forward and, at the same time, bring the right leg forward and sweep the opponent's right leg, bringing him to the ground.

Counter to a Rear Bear Hug

Marines use a counter to a rear bear hug when the opponent approaches from the rear and puts both of his arms around the Marine's body, trapping the Marine's arms to the sides. To execute the counter to a rear bear hug, Marines—

Step behind the opponent's right leg with the left foot, keeping both legs bent (almost in a squatting position). The left side of the body should be against the opponent's.

Pivot the hip, turning the body to the left and throwing the opponent back over the bent leg.

(reverse blank)

CHAPTER 7

GROUND FIGHTING

This chapter describes all techniques for a right-handed person. However, all techniques can be executed from either side.

In drawings, the Marine is depicted in woodland camouflage utilities; the opponent is depicted witlwut camouflage. In photographs, the Marine is depicted in woodland camouflage utilities; the opponent is depicted in desert camouflage utilities.

Marines should avoid being on the ground during a close combat situation because the battlefield may be covered with debris and there is an increased risk of injury. However, many close combat situations involve fighting on the ground. The priority in a ground fight is for Marines to get back on their feet as quickly as possible. In any ground fighting scenario, Marines will usually end up in one of four positions with the opponent. The offensive positions, in which Marines have a tactical advantage, are the guard and mount. The defensive positions, which are used as counters when the opponent has the tactical advantage, are the counter to the guard and the counter to the mount. Marines can also employ chokes during ground fighting to quickly end a fight.

1. Offensive Ground Fighting

Guard Position

Marines execute a guard position when the opponent is on top and the legs are wrapped around the opponent's legs. If the opponent is on top, he may try to choke the Marine, but the Marine still has the tactical advantage because the Marine is in a position to control the situation. To execute the guard position technique, Marines—

Trap the opponent's hands on the chest by crossing the hands on the chest so the forearms are resting on the opponent's forearms. Apply pressure with the elbows or forearms.

Strike the outside of the opponent's thigh with the cutting edge of the right heel. This causes the opponent to jerk to that side.

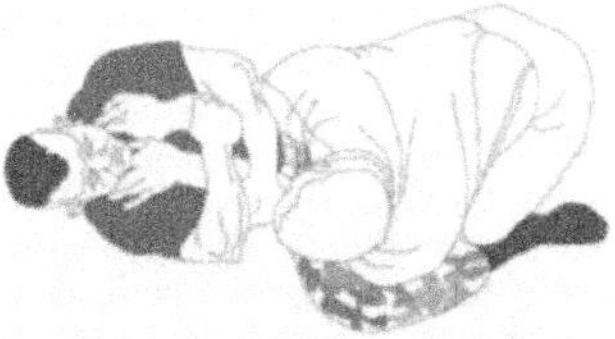

Move the head quickly to the left and swivel the hips to the right. At the same time, bring up both of the legs. Both legs are on the right side of the opponent's body.

Bring the right leg down, hooking the opponent's neck and head. Exert downward pressure to roll him over on his back. Grasp and maintain control of the opponent's left arm. Upon completion of the move, the Marine is sitting up with legs bent over the opponent while maintaining control of his arm.

Keep legs and knees bent. Maintain pressure against the opponent's neck with the back of the right foot and against his side with the left foot underneath his armpit. Squeeze the knees together, locking the opponent's arm.

Pull the opponent's arm straight up and fall back sharply, pulling his arm to the side in the direction of his little finger.

Return to standing.

Mount Position

Marines execute the mount if the opponent is lying on his back on the ground and the Marine is on top with legs wrapped around the opponent's body. This position is an offensive position because Marines are in a better position to control the opponent and to execute ground fighting techniques. To execute the mount position technique, Marines—

Grab the opponent's wrists or forearms with the hands. Hold them tightly against the torso.

Maintain control of the opponent's arms with the left arm and apply pressure to the opponent's brachial plexus (tie in) with the right hand.

Bring the left foot close to the opponent's right armpit and swing the right leg across the opponent' s head.

Keep the legs and knees bent. Maintain pressure against the opponent's neck with the back of the right foot and against his side with the left foot which is underneath his armpit. Squeeze the knees together, locking the opponent's arm.

Pull the opponent's arm straight up and fall back sharply, pulling his arm to the side in the direction of his little finger.

Return to standing.

2. Defensive Ground Fighting

Counter to the Guard

Marines use this technique if the opponent is lying on his back on the ground and the Marine is kneeling on the ground between the opponent's legs. To execute the counter to the guard position, Marines—

Strike the opponent's femoral nerve, located on the inside of the thigh, with the elbows. This forces the opponent to separate his legs.

Strike the opponent's groin with the right fist.

Hook the left arm underneath the opponent's right knee from the inside. Throw the opponent's leg over the head.

Duck the head quickly and move to the left.

Return to standing.

Counter to the Mount

Marines use this technique when they are lying on their back on the ground, the opponent is mounted on top of the Marine, and the opponent's legs are wrapped around the Marine. The opponent has the tactical advantage. To execute the counter to the mount position, Marines—

Grab the opponent's gear or clothing on his upper torso and pull him down close.

Use the right arm to hook the opponent's left arm, from the inside around the outside, above his elbow. With the right foot, hook the opponent's left leg or ankle.

Draw the elbow in to bend the opponent's elbow, bringing him down close. The arm must be hooked above the opponent's elbow in order to bend it.

Strike the opponent's side with the left hand.

Push the opponent over and roll him off to the right side.

Maintain control of the opponent's hooked arm and move to a kneeling position.

Move to a standing position while maintaining pressure on the opponent's arm and using the knee to apply pressure against the opponent's elbow.

3. Ground Fighting Chokes

WARNING

During training, never execute a choke at full force or full speed. Never hold a choke for more than 5 seconds.

The priority in ground fighting is for Marines to get back on their feet as quickly as possible. Sometimes, Marines can quickly end a ground fight by executing a choke on the opponent. When performed correctly, a choke can render an opponent unconscious in as little as 8 to 13 seconds. Chokes are easily performed regardless of size or gender. The chokes performed during ground fighting are the same as those performed while standing (see chap. 6).

Ground Fighting Front Choke

The ground fighting front choke is a blood choke performed most effectively from the mount position. The front choke employs the opponent's lapels or collar to execute the choke. To execute the ground fighting front choke, Marines—

Grab the opponent's right lapel with the right hand, making certain that the knuckles or the back of the hand are against the carotid artery on the right side of the opponent's neck.

Keep the right hand pressed against the opponent's carotid artery, reach under the right arm with the left hand, and grab the opponent's left lapel, forming an X with the wrists.

Keep the right hand pressed against the opponent's carotid artery and pull the opponent's left lapel to the left with the left hand.

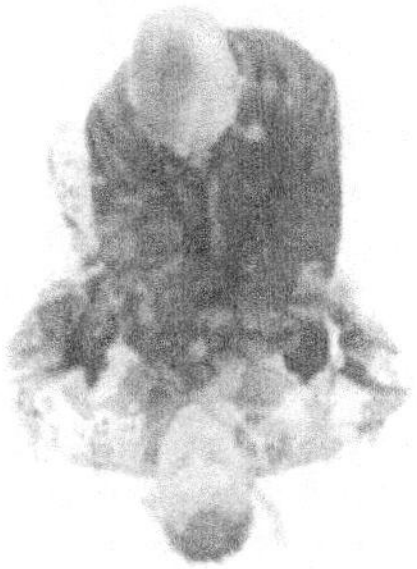

Ground Fighting Side Choke

The ground fighting side choke is a blood choke performed from the mount position. The ground fighting side choke is particularly effective when the opponent raises his arms and places them on the Marine's chest or throat. To execute the ground fighting side choke, Marines-

Use the left hand to parry the opponent's right arm inboard (to the inside of the opponent's reach). Bring the right arm underneath the opponent's arm and up around the front of his neck.

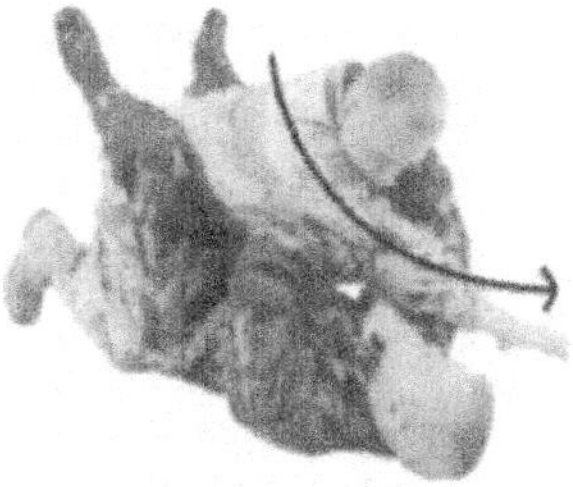

Keep the fingers extended, place the back of the forearm against the opponent's neck just below his ear and press the carotid artery on his neck. Depending on the Marine's or the opponent's body configuration, the back of the thumb or wrist may press against the opponent's neck.

Use the left hand to reach around the back of the opponent's neck and clasp the hands together.

Pull the opponent toward the chest. Use the forearm to exert pressure on the side of his neck. This is done by pulling the clasped hands toward the chest.

Ground Fighting Rear Choke

The ground fighting rear choke is a blood choke performed when Marines are behind the opponent. To execute the ground fighting rear choke, Marines—

Wrap the legs around the opponent with the left foot against the inside of his left thigh and the right foot against the inside of his right thigh.

Use the right arm to reach over the opponent's right shoulder and hook the bend of the arm around his neck.

Use the left hand to clasp both hands together.

Exert pressure with the biceps and forearms on both sides of the opponent's neck. This applies pressure to the opponent's carotid arteries. While maintaining pressure with the biceps and forearms, draw the opponent closer by drawing the right arm in. To increase the effectiveness of the choke, lean backward by arching the back and pulling the opponent back. At the same time, push the feet against the opponent's thighs.

Figure-Four Choke

A variation of the ground fighting rear choke is the figure-four choke. The figure-four choke allows Marines to gain more leverage on the rear choke. If the rear choke cannot be secured, the figure-four variation may be applied to increase the pressure of the choke on the opponent. To execute the figure-four variation of the rear choke, Marines—

Apply the ground fighting rear choke. The chest should be against the opponent's back.

Grasp the left biceps with the right hand and place the left hand against the back of the opponent's head.

Use the left hand to push the opponent's head forward and down.

Draw the right arm in while maintaining pressure with the biceps and forearm on both sides of the opponent's neck.

Increase the effectiveness of the choke by arching backward and pushing the feet against the opponent's thighs. Continue to pull the opponent back with the right arm while exerting pressure forward with the left hand against the opponent's head.

CHAPTER 8

NONLETHAL TECHNIQUES

This chapter describes all techniques for a riglit-handed person. However, all techniques can be executed from either side.

In drawings, the Marine is depicted in woodland camouflage utilities; the opponent is depicted without camouflage. In photographs, the Marine is depicted in woodland camouflage utilities; the opponent is depicted in desert camouflage utilities.

The Marine Corps' involvement in military operations other than war—e.g., humanitarian, peacekeeping, or evacuation missions—has greatly increased. These missions require skills that span the spectrum of conflict and support operations within a continuum of force. But the Corps' day-to-day existence also demands a responsible use of force. Nonlethal techniques are among the skills Marines use to apply a responsible use of force.

1. Unarmed Restraints and Manipulation

Marines operate within a continuum of force, particularly in support of peacekeeping or humanitarian types of missions. In these situations, Marines must act responsibly to handle a situation without resorting to deadly force. Unarmed restraints and manipulation techniques including joint manipulation, come-alongs, and takedowns can be used to control a subject without resorting to deadly force. Marines must train to become proficient in nonlethal techniques and to respond in a responsible manner. These techniques are referred to as compliance techniques, and they are applied in the third level in the continuum of force.

WARNING

During training, never apply the techniques for unarmed restraints and manipulation at full force or full speed. Use a slow and steady pressure to avoid injury.

Compliance Techniques

Compliance techniques are unarmed restraint and manipulation techniques used to physically force a subject or opponent to comply. Compliance can be achieved through the close combat techniques of-

Pain compliance using joint manipulation and pressure points. (Pain compliance is the initiation of pain to get compliance on the part of the subject.)

Come-along holds.

Principles of Joint Manipulation

Joint manipulation is used to initiate pain compliance and gain control of a subject. It involves the application of pressure on thejoints (elbow, wrist, shoulder, knee, ankle, and fingers). Pressure is applied in two ways:

In the direction in which the joint will not bend. For example, joints such as the knees and elbows only bend in one direction and when pressure is applied in the opposite direction, pain compliance can be achieved. Beyond the point where the joint stops naturally in its range of movement (i.e., it no longer bends).

Since each joint has a breaking point, Marines should apply slow steady pressure only until pain compliance is reached. Continued pressure will break the joint and may escalate the violence of the situation.

Joint manipulation also uses the principle of off-balancing. A subject can be better controlled when he is knocked off balance.

Wristlocks

A wristlock is a joint manipulation that can be applied in a number of ways to achieve pain compliance. The wrist rotates in a number of directions and will bend in a single direction until its movement stops naturally. In a wristlock, pressure is exerted beyond that point by bending or twisting the joint. A wristlock is executed when an opponent tries to grab Marines or is successful in grabbing Marines or their equipment. A wristlock can also be performed by Marines if they wish to initiate control of an opponent.

Basic Wristlock. A basic wristlock is executed when Marines grab the opponent's left hand with the right hand. To execute the basic wristlock, Marines—

Use the right hand to grab the opponent's left hand by placing the thumb on the back of the opponent's hand so that the Marine's knuckles are facing to the left.

Hook the fingers across the fleshy part of the opponent's palm below the thumb. The fingers are used to anchor the hand so leverage can be applied to twist and bend the joint.

Exert downward pressure with the thumb to bend the opponent'sjoint. Rotate the opponent's hand to the right to twist the joint.

Use the left hand to further control the opponent.

Step in to the opponent to keep the opponent's hand in close to the body to control him and provide more leverage on the wristlock.

Note: When executing the basic wristlock with the left hand, the Marine grabs the opponent's hand so that the Marine's knuckles are facing to the right, and then rotates and twists the opponent's hand to the left.

Reverse Wristlock. A reverse wristlock is executed when Marines grab the opponent's right hand with the right hand. To execute the reverse wristlock, Marines—

Place the right palm on the back of the opponent's right hand and wrap the fingers across the fleshy part of his palm below his little finger.

Twist the opponent's hand to the right while stepping in to place his hand against the chest. Apply downward pressure on the opponent's hand against the chest. Leave the opponent's hand on the chest to fully control the subject and to gain leverage.

Lean forward to use body weight to add additional pressure to the joint.

Use the left hand to further control the opponent.

Two-Handed Wristlock. Both hands can be used in the wristlock to maximize the leverage and pressure needed to bend and twist the joint. To execute the two-handed wristlock, Marines—

Place both thumbs on the back of the opponent's hand, thumbs crossed.

Hook the fingers of both hands around the fleshy part of the opponent's palm on both sides of his hand.

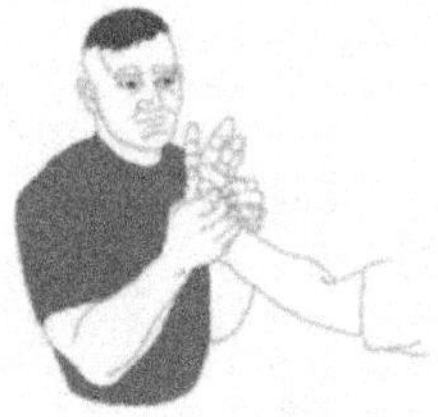

Step into the opponent and apply pressure downward on the back of his hand to bend the joint and rotate his wrist away from the body to twist the joint.

Enhanced Pain Compliance on Wristlock. Enhanced pain compliance techniques are applied in the third and fourth levels in the continuum of force. Additional pain be applied to a wristlock by-

Adding downward pressure to the elbow with the other hand or elbow by using the fingers to pull in on the opponent's radial nerve located on the inside of the forearm. When pressure is added to the opponent's radial nerve, his direction can be controlled.

Applying pressure against the opponent's finger joint to bend it in a direction it cannot bend (i.e., splitting the fingers).

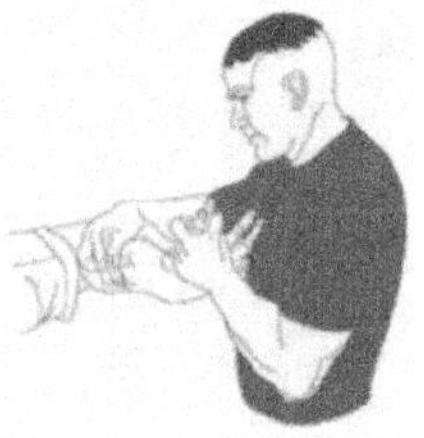

Come-Along Holds

Marines use a come-along hold to control and move an opponent.

Escort Position. A common come-along hold is the escort position. To execute the escort position, Marines—

Face the opponent. Use the left foot to step forward at a 45-degree angle. Turn to face the right side of the opponent.

Use the right hand to firmly grasp the opponent's right wrist. With the left hand, firmly grasp the opponent's right triceps.

Position the opponent's controlled arm diagonally across the torso, keeping his wrist against the right hip. The Marine should be standing to the right of and behind the opponent.

Note: This technique works well when escorting an opponent on either the right or left side. Take caution when escorting an opponent by ensuring his controlled hand is not in a position to grab the holstered weapon. The preferred escort position is from the Marine's left side, so that the opponent is kept further away from the weapon.

Wristlock Come-Along. To execute the wristlock come-along, Marines-

Use the left hand to execute a basic wristlock. Incorporate the right hand in a two-handed wristlock for more control.

Maintain pressure on the opponent's wrist with the right hand, step forward, and pivot around to stand next to the opponent.

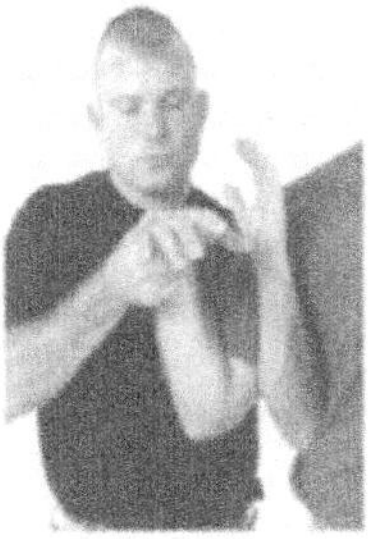

Release the left hand, quickly reach under the opponent's arm from behind, and grab his hand.

Use the left hand and apply downward pressure on the opponent's wrist.

Controlling Technique. The following controlling technique is used when an opponent grabs the Marine's wrist. To execute the technique, Marines—

Trap the opponent's hand with the palm of the other hand.

Rotate the opponent's trapped hand up and on his forearm while maintaining downward pressure on his trapped hand.

Apply downward pressure with both hands until the opponent is taken to the ground.

Armbars

An armbar is a joint manipulation in which pressure is applied on a locked elbow, just above the joint, in the direction the joint will not bend. An armbar has to be locked in quickly, but still requires a slow, steady pressure to gain compliance.

Basic Armbar. To execute a basic armbar, Marines-

Use the right hand to grab the opponent's right wrist.

Bring the left hand down on or above the opponent's elbow joint. To gain additional leverage, pivot to face the opponent.

Use the left hand to apply downward pressure on or above the opponent's elbow joint while pulling up on his wrist.

Armbar from a Wristlock. To execute an armbar from a wristlock, Marines-

Use the right hand to grab the opponent's right hand and execute a reverse wristlock.

Bring the left hand down on or above the opponent's elbow joint.

Use the left hand to apply downward pressure on or above the opponent's elbow joint while pulling up on his wrist.

Takedowns

A takedown is used to bring an opponent to the ground to further control him.

Takedown From a Wristlock Come-Along. To take the opponent to the ground from a wristlock come-along, Marines—

Use the right foot to push down on the opponent's calf or Achilles tendon.

Maintain control of the opponent's wrist and elbow and apply a slow, steady pressure to bring him to the ground.

Armbar to a Takedown. This technique is used to take a noncompliant opponent to the ground from an armbar. To execute the armbar takedown, Marines—

Use the right hand to execute a reverse wristlock.

Bring the left hand or forearm down on or above the opponent's elbow joint.

Pivot so the back is facing the opponent and, at the same time, lift the left elbow and slide the body so it is against the opponent, placing the armpit high above the opponent's elbow joint.

Lean back, placing the body weight on the opponent's arm until he complies or is taken to the ground.

Note: This technique may break the opponent's arm. Therefore, this technique should not be employed if the objective is a nonlethal takedown.

Wristlock Takedown. This technique is used to take a noncompliant opponent to the ground from a basic wristlock and to put him in a position where he can be handcuffed, if necessary. To execute the takedown, Marines-

Use the right hand to execute a basic wristlock. Incorporate the left hand in a two-handed wristlock.

Apply downward pressure on the wristlock, pivot on the ball of the right foot, and quickly turn to the right to take the opponent to the ground.

Continue to apply pressure on the wristjoint as the opponent lands on his back with his arm straight in the air.

Slide the left foot under the opponent's back, underneath his armpit.

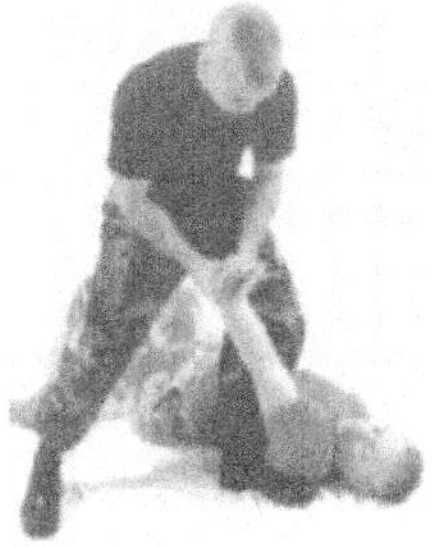

Apply pressure with the knee against the opponent's triceps while pulling back on his arm.

Continue applying pressure with the knee on the opponent's arm. Pivot and step around the opponent's arm to roll him on his stomach.

Kneel down with one knee on the opponent's back. The other knee is placed on the opponent's neck and shoulder on either side of his arm. Apply inward pressure with the knees to lock his arm in place.

Tell the opponent to put his other hand in the middle of his back. Bring the opponent's controlled hand to the center of his back.

Escort Position Takedown. This technique can be used to control a noncompliant opponent from the escort position. To execute the escort position takedown, Marines—

Lock the opponent's arm straight across the body while rotating his wrist away from the body.

Use the left hand or forearm and apply downward pressure above the opponent's elbow where the triceps meet.

Step back with the right foot and, keeping the opponent's hand controlled against the hip, pivot to the right while continuing to apply downward pressure on his arm to bring him to the ground.

Note: This technique works well when escorting a subject on the Marine's right or left side. When taking down a subject from the right side, step back and pivot to the left.

2. Nonlethal Baton

A baton or nightstick can be an effective compliance tool when used correctly. Batons can be used defensively (blocking), offensively (striking), and as a restraining device when needed. In the fourth level of the continuum of force (assaultive [bodily harm]), defensive tactics include baton or nightstick blocks and blows. Blows to the head or other bony parts of the body are considered deadly force. When deadly force is not authorized, Marines must be able to employ blocks, strikes, and restraints with the baton with the minimum of force.

Grips

One-Handed Grip. To execute the one-handed grip, Marines—

Use the right hand to grasp the lower end of the baton, about 2 inches from the end.

Wrap the thumb and index finger around the baton so they are touching one another. The grip on the baton should be firm, but natural.

Two-Handed Grip. To execute the two-handed grip, Marines-

Use the right hand to grasp the lower end of the baton, about 2 inches from the end. Wrap the thumb and index finger around the baton so they touch one another.

Use the left hand to grasp the upper end of the baton, palm down, about 2 inches from the end. The hands should be approximately 10 to 12 inches apart.

Stance and Method of Carry

The basic warrior stance serves as the foundation for initiating nonlethal baton techniques. The method of cany provides effective defensive positions with a wide range of options to control a combative opponent.

One-Handed Carry. To execute the one-handed carry, Marines—

Grip the baton using the one-handed grip.

Elevate the baton, with the gripping hand at a level between the belt and shoulder.

Keep the left hand in the position of the basic warrior stance.

Two-Handed Carry. This carry is effective for blocks. To execute the two-handed carry, Marines-

Grip the baton using the two-handed grip.

Elevate the baton, with the left hand higher than the right hand.

Orient the weapon toward the subject.

Movement

In a nonlethal confrontation, movement may be made to create distance between Marines and the opponent or to close the gap to control him. When facing an opponent, Marines move in a 45-degree angle to either side of the opponent. Moving at a 45-degree angle is the best way to both avoid an opponent's strike and to put Marines in the best position to control the opponent.

Blocking Techniques

One-handed blocks are used when carrying the baton in the one-handed carry. The same one-handed blocks used in combative stick techniques (see page 3-6) apply in nonlethal baton engagements. The blocks include blocks for a vertical strike, a forward strike, and a reverse strike. Because the baton is often carried with two hands, there are also two-handed blocks that are effectively used from this carry. Two-handed blocks are discussed in the following subparagraphs.

High Block. Marines execute a high block to deter a downward vertical attack directed at the head and shoulders. To execute the high block, Marines—

Raise the baton up to a level even with or above the head. The baton should be in a horizontal position to block the blow. The fingers of the left hand should be open and behind the baton.

Place the baton perpendicular to the opponent's striking surface to absorb the impact of the blow.

Bend the elbows slightly to help absorb the impact of the blow. The arms should give with the strike of the blow.

Low Block. Marines execute a low block to deter an upward vertical attack directed at the abdomen, groin, or torso. The opponent's blow can be delivered by a foot, knee, or fist. To execute the low block, Marines—

Lower the baton to a level even with or below the groin. The baton should be in a horizontal position to block the blow. The fingers of the left hand should be open and behind the baton.

Place the baton perpendicular to the opponent's striking surface to absorb the impact of the blow.

Bend the elbows slightly to help absorb the impact of the blow. The arms should give with the strike of the blow.

Right Block. Marines execute a right block to deter an attack directed at the head, neck, flank, or hip. The opponent's blow can be delivered by a foot, knee, fist, or elbow. To execute the right block, Marines-

Thrust the baton in a vertical position to the right side.

Pivot to the right by stepping with the left foot and pivoting off the ball of the right foot. Rotate the hips and shoulders into the direction of the block. The fingers of the left hand should be open and behind the baton.

Place the baton perpendicular to the opponent's striking surface to absorb the impact of the blow.

Bend the elbows slightly to help absorb the impact of the blow. The arms should give with the strike of the blow.

Left Block. Marines execute a left block to deter an attack directed at the head, neck, flank, or hip. The opponent's blow can be delivered by a foot, knee, fist, or elbow. To execute the left block, Marines—

Thrust the baton in a vertical position to the left side.

Pivot to the left by stepping with the right foot and pivoting off the ball of the left foot. Rotate the hips and shoulders into the direction of the block. The fingers of the left hand should be open and behind the baton.

Place the baton perpendicular to the opponent's striking surface to absorb the impact of the blow.

Bend the elbows slightly to help absorb the impact of the blow. The arms should give with the strike of the blow.

Middle Block. Marines execute either a left or right block to deter an attack directed at the face, throat, chest, or abdomen. To execute the middle block, Marines—

Thrust the baton in a vertical position straight out in front of the body. The fingers of the left hand should be open and behind the baton.

Place the baton perpendicular to the opponent's striking surface to absorb the impact of the blow. The baton should be held with the left hand forward of the right.

Bend the elbows slightly to help absorb the impact of the blow. The arms should give with the strike of the blow.

Restraining Technique

The strong-side armlock is used to restrain an opponent who is not compliant. To execute the strong-side armlock, Marines—

Use the right hand to run the baton up under the opponent's left armpit, parallel to the ground.

Use the right foot to step forward at a 45-degree angle to the left side of the opponent. The baton is across his forearm.

Use the right hand to drive the baton forward and up so the action bends the opponent's arm behind his back. At the same time, continue moving around him to get behind him.

Place the baton on the opponent's forearm with the thumb and/or fingers and apply pressure to his forearm. At the same time, grasp the other end of the baton with the left hand.

Pull up on the low end of the baton with the right hand. At the same time, push down on the top end of the baton with the left forearm, reaching around with the left hand to grasp the opponent's biceps or shoulder.

Continue exerting downward pressure with the left forearm while pulling back on the opponent's biceps with the left hand. This places the opponent in a position where he is controlled and can be moved.

Apply pressure with the foot against the bend in the opponent's leg above his calf. This lowers the opponent to the ground rather than throwing him to the ground and risking severe injury.

Striking Target Areas

Marines must avoid striking an opponent in the head, neck, or other bony parts with the baton because this is considered deadly force and can lead to serious bodily injury or death. Instead, the legs, arms, and buttocks are target areas that are considered nonlethal.

Legs. Primary targets are the thighs and lower legs. Avoid striking the knee and ankle joints because this can cause permanent damage.

Arms. Primary targets are the upper arms. Avoid striking the shoulder and elbow and wrist joints because this can cause permanent damage.

Buttocks. Primary targets are the buttocks. Avoid striking any other part of the torso, including the chest, rib cage, spine, tail bone, and groin because strikes to these areas can cause permanent damage or death.

One-Handed Striking Techniques

One-Handed Forward Strike. A forward strike follows either a horizontal line or a downward diagonal line using a forehand stroke. To execute the one-handed forward strike, Marines-

Stand facing the opponent with the baton carried in a one-handed cany.

Place the right hand palm up, swing the baton from right to left, and make contact with the opponent.

One-Handed Reverse Strike. A reverse strike follows either a horizontal line or a downward diagonal line using a backhand stroke. To execute the one-handed reverse strike, Marines—

Stand facing the opponent with the baton carried in a one-handed carry.

Bend the right arm and cross the arm to the left side of the body. The baton should be close to or over the left shoulder.

Place the right hand palm down, swing the baton from left to right, and make contact with the opponent.

Two-Handed Striking Techniques

Two-Handed Forward Strike. This strike is an effective follow-up to a middle block or left block. To execute the two-handed forward strike, Marines—

Pull back with the left hand while driving the right hand forward toward the opponent. The baton should be horizontal to the ground. Power is generated by stepping forward with the right foot and rotating the right hip and shoulder into the strike.

Contact the opponent with the end of the baton.

Two-Handed Reverse Strike. This strike is an effective follow-up to a middle block or left block. To execute the two-handed reverse strike, Marines—

Pull back with the right hand while driving the left hand forward toward the opponent. The baton should be horizontal to the ground. Power is generated by stepping forward slightly with the left foot and rotating the left hip and shoulder into the strike.

Contact the subject with the end of the baton.

Front Jab. This strike is effective for countering a frontal attack. It can also be executed as a quick poke to keep a subject away. To execute the front jab, Marines thrust both hands forward in a quick jab. The baton is held either horizontal to the ground or at a slight downward angle.

Rear Jab. This strike is effective for countering a bear hug from the rear. To execute the rear jab, Marines thrust both hands rearward in a quick jab. The baton is held either horizontal to the ground or at a slight downward angle.

APPENDIX A

PUGIL STICK TRAINING

A pugil stick is a training device used to simulate a rifle bayonet so that effective, but safe, training can be conducted to build proficiency of rifle bayonet techniques. Pugil stick training builds on the techniques used to throw punches. Pugil stick training is the only "full contact" training provided to Marines in the Close Combat Program. Pugil stick training teaches Marines to function when faced with stress and violence, and it prepares them to deliver a blow and take a blow. It also provides them with the physical and mental skills vital to success on the battlefield.

1. Pugil Stick Training

Design

A pugil stick consists of a stick wrapped in padding at both ends that can be gripped like a rifle. The pugil stick is approximately the same weight and length of an unloaded rifle with a bayonet attached.

Holding the Pugil Stick

The pugil stick is held in the same manner as the sen ice rifle. All movements come from the basic warrior stance. To hold the pugil stick correctly, Marines—

Use the right hand to grasp the lower end of the pugil stick overhanded.
Use the left hand to grasp the upper end of the pugil stick underhanded.
Use the right forearm to lock the lower end of the pugil stick against the hip.
Orient the blade end of the pugil stick toward the opponent.

Safety Equipment

The following safety equipment must be worn during any pugil stick bout.

Groin Protection. Groin protection protects the groin from an accidental blow. It should be pulled high around the waist to protect the groin area, with the concave portion against the body.

Flak Jacket. The flak jacket provides protection to the body. It is worn completely fastened.

Neck Roll. The neck roll prevents whiplash if Marines receive a blow to the head. The neck roll further supports the head and protects the neck from blows. The neck roll is worn above the flak jacket and below the helmet. The tied end of the neck roll faces the front.

Helmet. A regulation football helmet protects the face and head. The helmet must fit snugly and the chin straps must be adjusted and snapped.

Mouthpiece. The mouthpiece is worn on the upper teeth to protect the teeth.

Gloves. Marines wear gloves to protect the hands if pugil sticks do not have gloves built-in.

Pugil Stick Screening

Prior to pugil stick training, instructors must ask participating Marines the following questions. If a Marine answers "yes" to any question, he must see the corpsman or a competent medical authority who evaluates him and determines whether or not the he can participate in the training.

Are you on light duty?
Are you restricted to running shoes by a corpsman or a doctor?
Have you fought in a pugil stick bout within the last 7 days?
Have you received a blow to the head within the last 7 days?
Have you had a concussion within the last 6 months?
Have you had dental surgery within the last 24 hours?
Do you have stitches or staples on your body?
Have you had a shoulder or head injury within the last 5 years?
Are you taking a prescription drug?
Do you have an ear infection or current sinus infection?
Have you had a broken bone within the last 6 months?

Safety Personnel

The following safety personnel are required to conduct pugil stick training:

One close combat instructor must officiate the bout.
One close combat instructor trainer, commissioned officer, or staff noncommissioned officer must be in the training area to serve as range safety officer.
One corpsman must be in the training area.

Training Area

To prevent injury, Marines train on areas with soft footing (i.e., sand or grass). Training mats are not recommended because feet can stick to the mats, prohibiting movement or causing joint injuries by twisting a knee. Bouts should not take place on a hard surface area; e.g., a flight deck or parking lot. A boxing ring may be used to conduct pugil stick bouts; ring dimensions can vary as long as there is ample room to execute the techniques in the training area.

Second Impact Syndrome

Second Impact Syndrome occurs when a second blow to the head produces a second concussion that occurs within 1 week following a previous concussion (before recovery from the first concussion). Second Impact Syndrome causes rapid brain swelling and can cause death. Therefore, there must be 7 days between pugil stick bouts to reduce the risk of severe injury resulting from Second Impact Syndrome. The 7 day separation between pugil stick bouts significantly reduces the possibility of injury, particularly in someone who may have suffered a brain injury or concussion but shows no symptoms.

Any Marine who experiences headaches or the following symptoms after training must be examined by appropriate medical personnel:

Blurred vision.
Ringing in the ears.
Dilation of the pupils.
Slurred speech.
Bleeding from ears or mouth.
Swelling in head or neck area.
Any unnatural discoloration in head or neck area.

The Marine should not be allowed to participate in pugil stick training or any other activity where a heavy blow might be sustained for a minimum of 7 days after the headache or other symptoms have subsided.

Safety Measures

The following safety measures must be followed:

Gear must be worn properly throughout training.
Contact lenses or glasses will not be worn.
False teeth will be removed from the mouth.
Nothing will be worn around the neck except the neck roll.

Competition among groups of Marines is authorized as long as it does not overshadow training objectives or compromise safety procedures.

Safety and proper techniques are paramount.

Safety is more important than competition.

2. General Rules and Regulations Governing Pugil Stick Bouts

Instructors and Support Personnel Requirements

At a minimum, one close combat instructor observes a pugil stick bout. For safety purposes, it is better to have two instructors judging a bout because each instructor can fully observe each of the fighters. The best position for observation is to the right of a fighter. This allows the instructor to see the fighter's facial expression and body movement. The instructor's position must not interfere with the fight.

In addition, one commissioned or staff noncommissioned officer (to serve as range safety officer) and one corpsman will be in the training area.

Prior to the Bout

Prior to the bout, Marines—

Are paired according to height, weight, and gender.

Wear the proper safety equipment.

Once Marines have properly donned the safety equipment, they wait for instructions from the close combat instructor. On the command of the close combat instructor, a pair of Marines will enter the ring. Once in the ring, the close combat instructor inspects each Marine for the proper safety equipment.

Note: There should be enough gear so that when two Marines are training, two other Marines can be donning safety equipment for the next bout.

During the Bout

The bout begins when the close combat instructor blows the whistle. All strikes are directed above the waist. Upon hearing a whistle blast, all fighting immediately ceases.

Stopping a Bout

There arc two reasons for stopping a bout: delivery of a scoring blow or an unsafe condition. The close combat instructor trainer, close combat instructor, or the range safety officer may stop the bout at any time an unsafe condition is observed.

Scoring Blow. A scoring blow is an offensive technique delivered to a vulnerable area of an opponent with sufficient force and precision to be considered as a disabling or killing blow. Scoring blows are notjudged solely on the degree of force with which a blow is delivered, but on the accuracy and techniques employed. A scoring blow is defined as—

A straight thrust with the blade end of the weapon (red end of the pugil stick) to the opponent's face mask or throat.

A slash to the side of the opponent's helmet (below the ear) or neck with the red end of the pugil stick.

A heavy blow to the opponent's head with an authorized technique (i.e., buttstrokc, smash) using the butt (black end) of the pugil stick.

When a scoring blow is delivered, the close combat instructor blows the whistle to stop the bout.

Unsafe Condition. The bout will be stopped as soon as an unsafe condition exists. An unsafe condition exists when a Marine is unable to defend himself, loses his balance and falls down on one or two knees or falls down completely, shows instability (e.g., buckling at the knees), loses muscular tension in his neck and his head snaps back or to one side, or appears disoriented. An unsafe condition also exists if a Marine lets go of one end of the pugil stick, the equipment (e.g., helmet, neck roll) falls off, or a Marine fails to use the proper techniques.

If any of these conditions occur—

The close combat instructor trainer, close combat instructor, range safety officer, or anyone supervising training stops the bout and separates the two Marines.

The corpsman evaluates the possibly injured Marine to determine if the fight can continue. The corpsman—

- Ensures that he is alert and responsive, both verbally and physically.
- Talks to him to see if he is coherent.
- Makes certain he comprehends and replies to verbal questions.
- Makes certain his speech is not slurred.
- Checks his physical signs.
- Ensures his eyes are focused and not dazed or glazed.
- Ensures his legs are not wobbly or shaky.

The pugil stick is not used as a baseball bat. The use of unauthorized techniques will result in expulsion from the bout. Only techniques taught in bayonet training are authorized. These techniques include-

Straight thrust.
Buttstroke (horizontal and vertical).
Smash.
Slash.
Parry.
Blocks (low, high, left, right).

After the Bout

A third whistle blast is used prior to identifying the Marine who delivered the killing blow.

A fourth whistle blast is used prior to demonstrating the killing technique used by the Marine who won the bout.

3. Directions for Making Pugil Sticks

Close combat instructor trainers are taught to make pugil sticks to standard specifications at school. It is recommended that all pugil sticks be made under the supervision of a close combat instructor trainer. Until the Marine Corps acquires a standardized pugil stick that units can order through the supply system, instructors use the following information and instructions to make pugil sticks locally.

Supplies

The following supplies are needed to make one pugil stick:

One 45" by 2" circular, oak, wood dowel.
Four 9" wide by 27" long foam pad pieces to make two inner handguards of double thickness.
Four 11" wide by 19" long foam pad pieces to make two outer handguards of double thickness.
Canvas material to cover inner and outer handguards.
One foam pad cut 1/4" thick by 7" wide by 13" long to make the centerguard.
One 8" wide by 14"long piece of canvas material to make a pugil stick sock.
Canvas material to make two pugil stick end caps. Ideally, two different color fabrics should be used to simulate the bayonet and buttstock ends of a rifle. Common colors for end caps are black and red. Each end cap is a different color to help the officiating certified close combat instructor determine the scoring blow during the pugil stick bout.
Two 1/4" thick by 11" wide by 54" long foam pad pieces to make two ends.
Two foam pads cut 1/4" thick by 3" wide by 14 1/2" long to make two end plug inserts.
One can of 3M Photo MountQ spray adhesive (NSN 8040-01-120-4009)to adhere foam padding to the stick. (Spray adhesive is available from self-service.)
One roll of duct tape or riggers tape (NSN 7510-00-266-5016) to protect foam pad and

canvas from wear and to reinforce gluing action of spray adhesive. (Duct tape or riggers tape is available from self-service.)

One roll para-cord or 5-50 cord to tie both the inner and outer handguards together.

One roll of thread strong enough to sew canvas material together.

Sixteen grommets, one for each of the four corners of the inner and outer handguards.

One foam sleeping mat (NSN 8465-01-109-3369). Use unserviceable foam sleeping mats obtained from Defense Reutilization and Marketing Office.

Tools

The following tools are needed to make a pugil stick:

Tape measure, ruler, or yardstick.
Scissors.
Grommet machine.
Knife.
Sewing machine.

Directions

To make a pugil stick, Marines perform the following steps.

Preparing Materials

Step 1. Cut foam sleeping mat into the dimensions listed above. Cut mats to double thickness.

Step 2. Use spray adhesive to adhere the two 9" by 27" pieces of foam together. Use the spray adhesive to adhere the two 11" by 19" pieces of foam together. Let adhesive dry before using the double-thick foam for handguards.

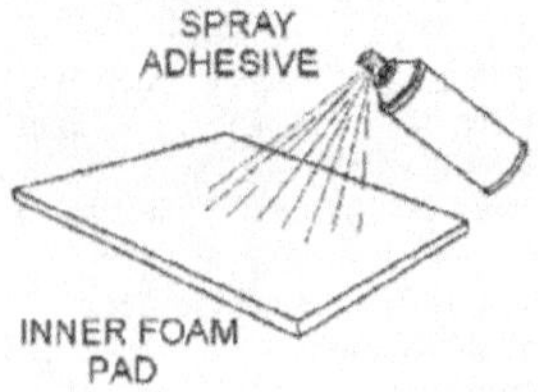

Step 3. Cut canvas material to make two coverings for the inner handguards. Cut material to make two coverings for the outer handguards. Two inner handguards and two outer handguards are required for each pugil stick.

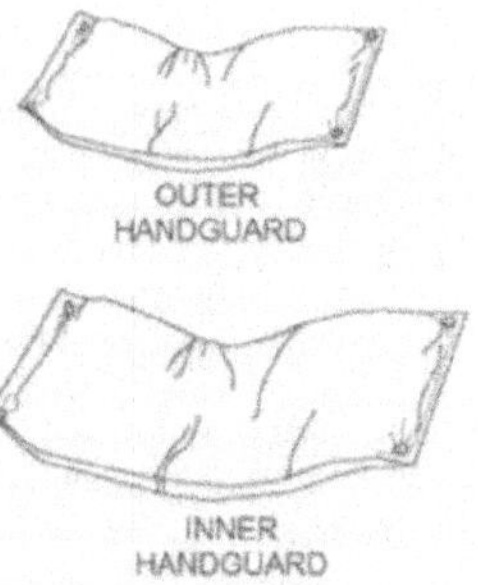

Step 4. Use a sewing machine to make a sleeve out of the 9" by 27" canvas. Sew down three sides. Insert foam pad inside the canvas sleeve and then sew down the fourth side. Use the grommet machine to create holes reinforced with metal grommets in each of the four corners. Once finished, the inner handguards are complete.

Note: It is recommended that the Fabric Repair Shop be used to sew canvas sleeves and apply grommets.

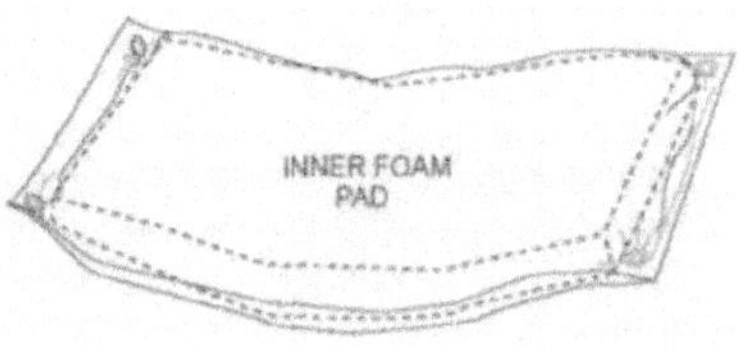

Step 5. Repeat step 4 to make the other inner handguard and both outer handguards.

Constructing Ends, End Plugs, and End Caps

Step 1. Apply spray adhesive on approximately 8 inches of one end of the dowel. This adhesive serves to glue the foam pad to the wooden dowel.

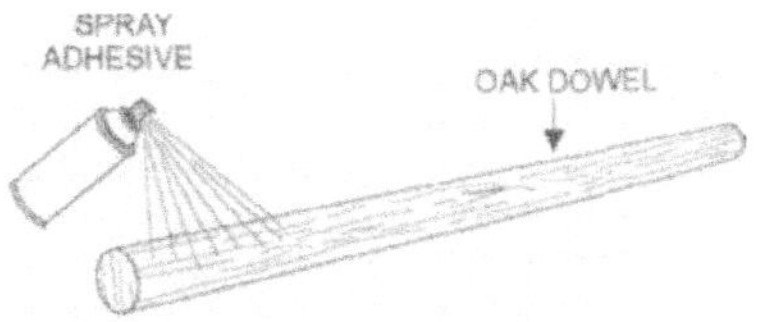

Step 2. Apply spray adhesive on one side of the 11" by 54" foam pad. This causes the foam pad to stick to itself when it is rolled around the end of the wooden dowel. Ensure 3 inches of the rolled foam pad extends past the wooden dowel for stability.

Step 3. Apply spray adhesive on one side of 3" by 14 1/2" foam pad. Roll pad into a tight cylinder.

Step 4. Apply spray adhesive on the exposed end of the dowel and fill the hole in the end with the end plug. The end plug prevents the wooden dowel from protruding beyond the edge of the rolled foam padding. This is an important safety precaution to prevent injury.

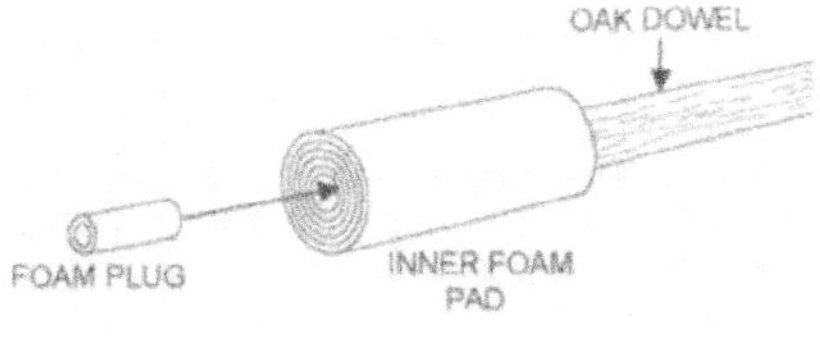

Step 5. Use duct tape or riggers tape to secure and reinforce the foam pad once the end plug is inserted in the end.

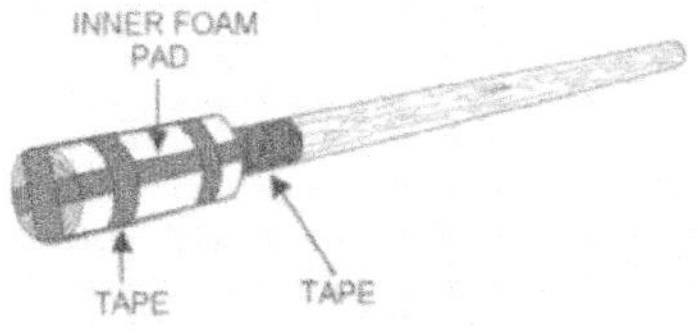

Step 6. Measure the dimensions of the ends (typically, the length of the cylinder should be 12 inches long to allow for taping). Cut and sew canvas material in the shape of a cylinder with one open end. Once complete, this becomes the end cap. The end cap is a colored sleeve that covers the rolled and taped foam pad at each end of the pugil stick.

Note: It is recommended that the Fabric Repair Shop be used to make the end cap.

Step 7. Slide one of the end caps over the rolled foam pad and tape it to the dowel.

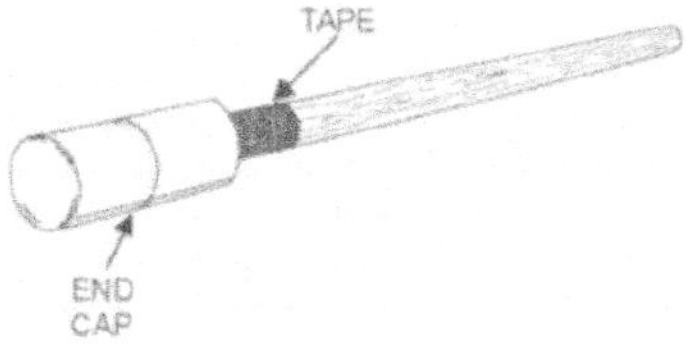

Building the Centerguard

Step 1. Spray one side of the 7" by 13" foam pad with spray adhesive and wrap it tightly around the center of the dowel.

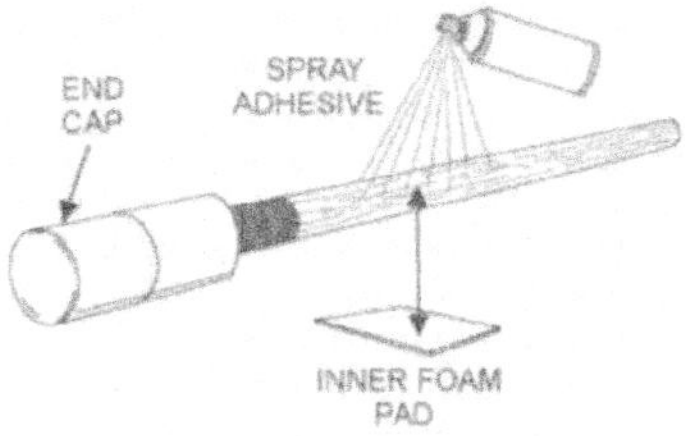

Step 2. Wrap duct tape or riggers tape around the foam pad to secure.

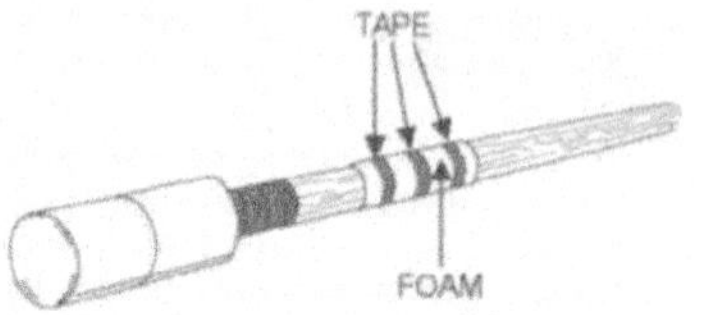

Step 3. Use a sewing machine to make a tubular pugil stick sock out of the 8" by 14" piece of canvas. Slide the pugil stick sock over the rolled foam pad and secure it with the duct tape or riggers tape. Duct tape or riggers tape protects the edges of the centerguard and reinforces the gluing effects of the spray adhesive.

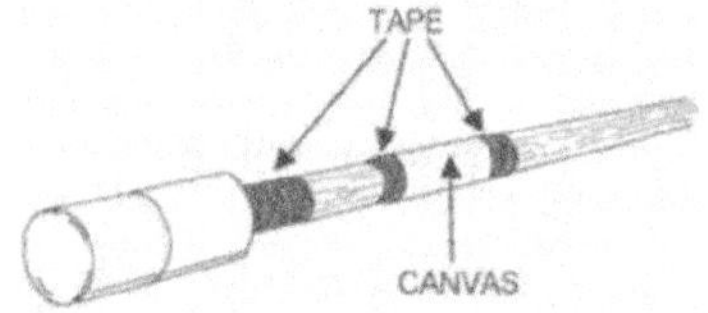

Note: It is recommended that the Fabric Repair Shop be used to make the pugil stick sock.

Finishing the Opposite End Cap

After one end cap is complete and the centerguard is finished, then repeat steps 1 through 7 in Constructing Ends, End Plugs, and End Caps for the other end of the pugil stick. Ensure that the opposite color end cap is used.

Fastening Handguards

Step 1. Place the inner handguards (longer piece) around the dowel. The inner handguard is placed on first because it protects the wrist. One inner handguard is needed for the left hand and one for the right hand.

Step 2. Fold the inner handguard around the pugil stick and run the para-cord or 5-50 cord through the metal grommets.

Step 3. Tighten down the corner ends of the inner handguards by pulling on the para-cord or 5-50 cord.

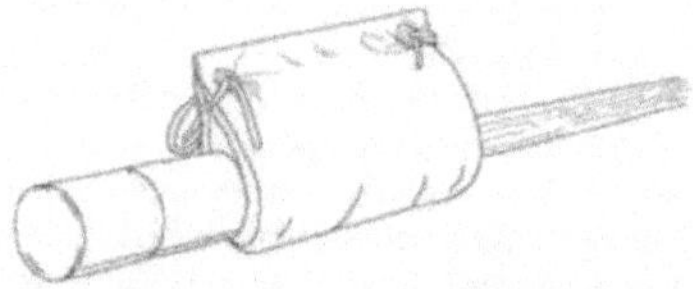

Step 4. Tie off the cord with a square knot above the metal grommets.

Step 5. Place the outer handguards (shorter piece) over the inner handguards (longer piece). Repeat steps 1-4 for the other handguard. The outer handguards provide additional protection to the knuckles and fingers and provides additional stability for the inner handguards.

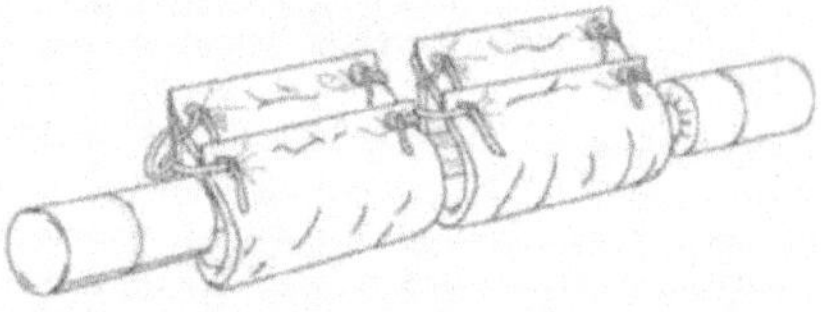

Serviceability Inspection

Prior to any training, all pugil sticks are inspected to ensure they are fit for use. If any of the following conditions exist, the pugil stick is unsafe and will not be used:

Stick protrudes through the end caps.
Frayed or worn edges exist on canvas surfaces.
The stick can be felt through the pads or the padding is too soft.
End caps or handguards are not securely attached.

(reverse blank)

APPENDIX B

SAFETY PRECAUTIONS DURING TRAINING

1. General Safety Precautions

When training close combat techniques, certain safety precautions must be adhered to to prevent injuries.

Most training should be conducted on a training area with soft footing such as a sandy or grassy area. If training mats are available, they should be used. A hard surface area is not appropriate for close combat training.

All techniques should be executed slowly at first. Marines can increase the speed of execution as they become more proficient. Marine-on-Marine training that requires contact (chokes, throws, ground fighting, and unarmed restraints and manipulation) should not be executed at full force or full speed.

If a technique is applied to the point that a Marine is uncomfortable, the Marine must "tap out." This indicates immediate release of the pressure being applied or to immediately stop the technique. The Marine "taps out" by firmly tapping his hand several times on any part of the opponent's body that will get his attention or by saying *stop*.

Second Impact Syndrome occurs when a second concussion develops within hours, days, or weeks following a prior concussion (and before recovery from the first concussion). Second Impact Syndrome causes rapid brain swelling and can cause death. Marines who experience headaches or other symptoms following training must be examined by appropriate medical personnel. These symptoms can include, but are not limited to, blurred vision, ringing in the ears, dialation of the pupils, bleeding from the ears or mouth, slurred speech, swelling in head or neck area, or any unnatural discoloration of head or neck. They should not be allowed to participate in pugil stick training or any other activity where a heavy blow might be sustained for a minimum of 7 days after the symptoms have subsided.

2. Safety Precautions for Individual Techniques

Falls

When training Marines to fall, they should progress from the ground, to a kneeling or squatting position, and then to a standing position. This ensures they are comfortable and understand the technique before progressing to executing falls from a higher profile. This instructional technique vastly reduces the risk of injury.

Strikes and Punches

When training Marines to strike and punch, they begin by executing the techniques "in the air." As they become more proficient, they execute strikes on equipment (when available) such as an air shield or a heavy bag. At no time should they be permitted to execute strikes on another student.

Chokes and Ground Fighting

Chokes

When training Marines to execute chokes, they will not apply pressure to the opponent's throat during training because the trachea and windpipe can be crushed. During training, Marines should practice the proper procedures for blood chokes. They should never execute air chokes. At no time during training should a choke be applied at full force or full speed. No choke should be held for longer than 5 seconds.

Unarmed Restraints and Manipulation

When training Marines to execute unarmed restraints and manipulation techniques, they utilize slow and steady pressure. Never apply these techniques at full force or at full speed.

Pugil Stick Training

A blow to the head during training will bruise the brain. A second blow to the head can cause death. This is known as the Second Impact Syndrome. There must be 7 days between pugil stick training to prevent injury or death.

*TC 21-24
TRAINING CIRCULAR
No. 21-24
HEADQUARTERS
DEPARTMENT OF THE ARMY
WASHINGTON, DC, 10 September 1997

RAPPELLING

CONTENTS

DISTRIBUTION RESTRICTION: Approved for public release; distribution is unlimited.

*This publication supersedes TC 21-24, 24 September 1991.

Page

PREFACE

This circular provides basic rappelling techniques to soldiers and leaders for the conduct of rappelling operations. It serves as the primary reference for both resident and nonresident instruction presented to cadets, officer candidates, and both commissioned and noncommissioned officers. This circular also discusses several advanced techniques dealing with infiltration and exfiltration.

Safety is always the most important consideration when conducting training. This circular provides guidelines for commanders to conduct operations safely. Commanders at all levels must analyze the complete training event to determine the degree of risk involved to men and or equipment (Appendix A). After determining the risks, risk reduction options or controls will be integrated into the training activity. These options or controls may range from a safety briefing or providing additional safety resources, to selecting other means of accomplishing the mission.

Rappelling, FRIES, and SPIES operations are inherently dangerous; therefore, the safety notes and considerations presented in this circular are the minimum acceptable standards.

The proponent of this publication (except for Chapter 5) is the US Army Infantry School. Submit changes for improving this publication on DA Form 2028 and forward it to the Commandant, US Army Infantry School, ATTN: ATSH-RB, Fort Benning, Georgia 31905-5593 (e-mail address: IKNERH@BENNING-EMH2.ARMY.MIL).

The proponent for Chapter 5, Fast-Rope Insertion/Extraction System, is the US Army Special Operations Command. Submit changes or comments for Chapter 5 on DA Form 2028 and forward it to the Commander, US Army Special Operations Command, ATTN: AOOP-TRS, Fort Bragg, North Carolina 28307-5000.

Unless this publication states otherwise, masculine nouns and pronouns do not refer exclusively to men.

CHAPTER 1

TOWER RAPPELLING

The introduction to rappelling is taught on a static tower (Figure 1-1). Using the building block approach to training, soldiers systematically progress to more demanding platforms, including taller static towers, and finally to helicopters. The static tower used may vary in size and height from 34 to 90 feet. The concept of learning the basic rappelling techniques before helicopter operations does not vary. (Units may conduct ground training and wall-side tower rappelling without a rappel master. In this case, the unit commander would be responsible for the safety of the rappelling training.)

Figure 1-1. Rappel tower.

Note: Each static rappel tower should resemble the structure or aircraft being trained to. The Army's standard design for building static rappel towers can be obtained through the Plans and Support Branch of the post Engineers, Building 75, Fort Benning, Georgia 31905 (commercial 706 545-3307 or DSN 835-3307). Strict adherence to the design specifications is required for safety considerations. Rappel towers should be inspected annually by post safety or post engineers.

Section I. PERSONNEL

This section discusses the personnel involved in training rappelling and their duties and responsibilities.

1-1. RAPPEL MASTER

The proponent for accreditation, evaluation, and information for Army rappel master courses is Directorate of Operations and Training, US Army Infantry School, ATTN: ATSH-OTT-F, Fort Benning, Georgia 31905-5593.

a. **Duties and Responsibilities.** The rappel master is responsible for the safety of rappellers. He ensures that all equipment (installation, unit, and personal property) is serviceable. He personally supervises the rappelling operation.

b. **Qualification.** Rappel master qualification is awarded only after the successful completion of a TRADOC accredited rappel master course, which includes the following subjects:

- Duties and responsibilities of a rappel master.
- Safety SOP, regulations, and references.
- Construction of a deployment bag.
- Conduct of an equipment rappel off the rappel tower.
- Conduct of a lock-in.
- Talking a rappeller through completion of a rappel.
- Conduct of ground training.
- Inspecting for proper hook-ups.
- Inspecting and maintaining equipment.
- Inspecting and maintaining snaplinks.
- Inspecting and maintaining rappelling gloves.
- Inspecting and maintaining rappel ropes.
- Identifying the rappel capabilities of aircraft used.
- Controlling rappels from UH-1H or UH-60 aircraft.
- Tying knots (square, bowline, half-hitch, Prusik), safety lines, and rappel seats (Swiss seat, Australian seat).
- Inspecting a rappel seat.
- Aircraft rigging for rappelling operations.
- Aircraft command and control.

c. **Proficiency Maintenance.** To remain current, rappel masters must execute their duties in a tactical or training exercise once every six months. If rappel masters do not execute their duties once every six months, they must take a refresher class taught by a current rappel master. The refresher class includes the subjects listed in the rappel master qualifications paragraph (1-1b).

1-2. RAPPEL SAFETY OFFICER

The RSO is an air assault-qualified or ranger-qualified SFC or above who serves as the OIC for the conduct of rappel operations.

a. The RSO is responsible overall for the safety of all rappellers and ensures that all safety precautions are followed.

b. The RSO briefs VIPs, visitors, and inspecting authorities on training, safety requirements, and layout of training areas.

1-3. RAPPEL LANE NCO

a. **Duties and Responsibilities.** Safety is the rappel lane NCO's number one priority. The rappel lane NCO—

- Ensures proper safety procedures are followed.
- Ensures proper hookup once directed to a rope station.
- Issues commands and maintains eye contact with the rappeller at all times.

b. **Qualifications.** The rappel lane NCO holds the rank of corporal or above, is ranger- or air assault-qualified, and is selected by the commander. Each tower rappel lane must have a qualified rappel lane NCO.

c. **Training.** The rappel lane NCO must also train on the following subjects:

- Responsibilities and safety requirements.
- Inspection and maintenance of equipment.
- Identification of satisfactory anchor points.
- Identification of safe and unsafe hookups.
- Establishment of a rappel point.
- Inspection of a rappel seat.
- Coaching techniques.
- Rappelling procedures.
- Emergency procedures.
- Belay control procedures.

d. **Participation.** The rappel lane NCO must participate in at least seven rappel operations: three as a rappeller; two as an assistant rappel lane NCO; and two performing the duties of a rappel lane NCO under the supervision of a qualified rappel master.

e. **Proficiency Maintenance.** If a rappel lane NCO has not conducted his duties within the last six months, he must complete the training listed in paragraph 1-3c under the supervision of a current rappel master.

1-4. RAPPELLER

Rappel qualification requirements apply to the individual rappeller. Participants in tower rappel training must complete the following requirements under the supervision of a rappel master. The unit commander ensures that personnel successfully complete these requirements before beginning aircraft rappel training.

- Identify all rappelling equipment.
- Demonstrate the construction and attachment of the rappel seat and the rappel rope to the seat.
- Identify unsafe attachments, equipment, rope connections, and seat construction.
- Define terms used in rappelling operations.
- Identify knots used in rappel operations.
- Understand and demonstrate rappel commands.
- Demonstrate rappelling positions.
- Demonstrate belaying procedures.
- Exhibit satisfactory performance from a rappel tower of at least 34 feet in height (two rappels with equipment and weapon, two without equipment and weapon). Two rappels are conducted from the free side of the tower (no wall).
- Demonstrate the ability to lock in.

1-5. BELAYER

Belay requirements are a subtask of basic rappel requirements. Soldiers must know how to belay before conducting rappelling training. The belayer—

- Assumes a position at the base of the lane about one pace away from the tower area.
- Ensures that the rappel ropes are even with the ground during tower rappels.
- Loosely holds the rappel rope with both hands so as not to interfere with the rappeller but still be able to stop the rappeller should he fall.
- Immediately stops the rappeller by pulling downward on the rappel ropes if the rappeller shouts "falling" or loses control of his brake hand during descent.
- Does not wear gloves to ensure a firm grip on the rappelling rope.
- Watches the rappeller at all times, and maintains constant voice or visual contact.
- Wears a helmet to prevent injuries from falling debris.

1-6. BELAY SAFETY

The belay safety must be ranger- or air assault-qualified. He ensures belay personnel are performing their duties properly. Rappel training requires one belay safety for each two rappel stations.

Section II. PREOPERATIONS BRIEFINGS AND SAFETY PROCEDURES

The rappel master ensures participants have a basic understanding of requirements and safety procedures before conducting training.

1-7. SAFETY

The following personnel and equipment must be present during static tower training.

- Two military rappel ropes for each rappel station.
- One safety officer.
- One rappel master for each rappel site.
- One rappel lane NCO per rappel station.
- One medic with medical kit and backboard.
- One safety or medical evacuation vehicle with driver.
- One belayer for each rope station. Rappellers alternate stations.
- One belay safety for each two rappel stations (four ropes).

1-8. SAFETY BRIEFING

As in all training, a safety briefing precedes rappel operations. The rappel master briefs all personnel on safety to include the following instructions.

a. Each rappeller ensures loose clothing and equipment are secured.

b. Rappel seats are tied by the soldier and inspected by the rappel master before climbing the tower. Rappel seats are removed upon completion of every rappel, retied, and reinspected by a qualified rappel master or rappel lane NCO before subsequent rappels.

c. Rappellers climb the tower only when directed by the rappel master or rappel lane NCO.

d. Rappellers stay in the center of the tower until instructed to move to a rappel point.

e. No more than three personnel are behind each rappelling point.

f. If using a troop ladder, only three soldiers are on the ladder at one time. Soldiers do not climb the ladder until told to do so by a rappel master.

g. All rappel masters, rappel lane NCOs, instructors, and anyone else standing near the edge of the top of the tower must wear a restraining strap or safety rope. The strap or rope must be attached to an anchor point.

h. No one should lean or sit on the railing or banisters of the tower.

i. No one is allowed within 3 feet of the edge of the tower without being secured.

j. When attaching the rappel rope to the snaplink, rappellers pull the slack towards the anchor point. The rappel master or rappel lane NCO physically check each hookup.

k. All personnel weighing more than 200 pounds will conduct a standard hookup rappel to determine if they require a friction hookup. A friction hookup is created by placing an additional two ropes in the gate of the snaplink (for a total of six ropes in the snaplink).

l. Combat equipment is positioned on the rappeller so that it does not interfere with the brake hand. The weapon must be slung diagonally across the back with the muzzle pointing down, and on the opposite side of the brake hand.

m. Heavy duty gloves are required for all rappel training.

n. Kevlar helmets with chin straps fastened are worn during tower rappel training.

o. While on the tower, the rappeller maintains eye contact with the rappel master or rappel lane NCO and receives all commands from them.

p. The rappeller ensures that he has a belayer on his rope.

q. The belayer does not wear gloves and keeps both hands on the rope at all times. He also faces the rappeller at all times.

r. All tower rappelling is performed with a double strand of rope.

s. No running is allowed on the tower.

t. No smoking or eating is allowed near the tower.

u. All participants who are unable to rappel, lack confidence, or refuse to rappel are reported to the rappel master or OIC. These participants are immediately removed from the training area.

v. The RSO and rappel master must be aware of overconfidence and carelessness of some rappellers. The rappel master ensures all personnel are tower-qualified before beginning aircraft rappel training.

1-9. TOWER SAFETY AND PREPARATION

The rappel master is in charge of the tower. He conducts a visual and physical inspection of every item of equipment to include the structural lumber and timber, the ladder, the platform floor, and all anchor points.

a. The static tower will not be used during thunderstorms or excessively high winds. If ice is present, or if the platform is slick from rain, rappelling will be delayed until conditions are safe.

A B

Figure 1-2. Example of rappel tower anchor points.

b. All rope stations are rigged with two anchor points (Figure 1-2A, B, C). The first anchor point is a middle-of-the-rope knot, and the second is an end-of-the-rope anchor knot. The rappel master removes all the slack between the knots to create equal tension on the anchor points. He ensures that no less than 10 feet of rope is on the ground during static rappelling.

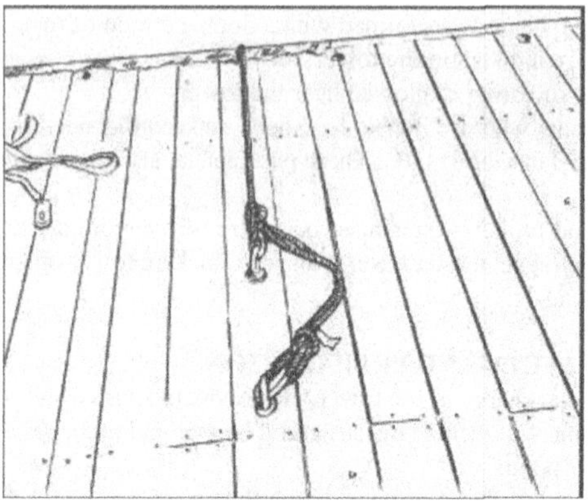

C

Figure 1-2. Example of rappel tower anchor points (continued).

1-10. RAPPELLER PREPARATION

Before conducting a rappel, each rappeller must prepare his individual clothing and equipment.

> **WARNING**
>
> **Failure to properly prepare rappellers could result in bodily injury or damage to equipment.**

a. Secure shirt tails, loose clothing, equipment, straps, and long hair.

b. Wear a helmet during rappelling. Properly fasten all straps and ensure the helmet is in serviceable condition.

c. Wear heavy leather workman's gloves.

d. Wear identification tags.

e. When rappelling with equipment, LBE or LBV should be unfastened in the front or fastened loosely behind the back of the rappeller. The rucksack should be worn high and tight on the back of the rappeller to allow the brake hand to reach the small of the back. Rucksack adjustment straps will be tied across the chest or tucked away.

f. Sling the weapon diagonally across the back with the muzzle down. Ensure the muzzle is on the guide hand side and the stock is towards the brake hand.

Note: Soldiers rappelling with equipment in excess of 50 pounds may want to consider using a friction hookup.

Section III. RAPPELLING PROCEDURES

This section discusses procedures used in tower rappelling.

1-11. SEAT-HIP RAPPEL

When using the seat-hip rappel, friction is created by a snaplink that is inserted in a sling rope seat and fastened to the rappeller. This method provides a faster and more controlled descent than other methods. Wear gloves to prevent rope burns. An alternate technique is to insert a second snaplink into the first snaplink (attached to rappel seat) and run the rope through the second snaplink. This allows easier disengagement from the rappel rope without running the entire rope through the first snaplink. To disengage from the rappel rope using the alternate technique, release the tension from the rope by opening the gate of the first snaplink and removing the second snaplink (with the rope attached).

a. The rappel seat is constructed as follows (Figure 1-3A through 1-3T):

(1) Place the midpoint (center) of the length of the sling rope on the hip opposite the brake hand (the brake hand is the strong hand) (Figure 1-3A, B, C).

(2) Bring the sling rope around the waist above the hip bone. Tie a double overhand knot over the navel (Figure 1-3D, E, F, G, H).

(3) Let the two free ends of the sling rope fall to the ground in front (Figure 1-3I).

(4) Bring the two free ends of the sling rope down between the legs and up over the buttocks. Ensure that the two free ends do not cross (Figure 1-3J).

(5) Pass the ends of the ropes over the rope that is tied around the waist at the two points above the center of the two rear seat pockets (Figure 1-3K).

(6) Grab the free end of the rope that is on the left side of the body with the left hand, and the free end of the rope that is on the right side of the body with the right hand.

(7) Squat down and simultaneously pull on both running ends of the ropes and stand up. This will tighten the seat.

(8) Take the two running ends of the rope down and back over the waist rope from the inside. Bring the running ends back under the ropes that are going across the buttocks (Figure 1-3L).

(9) Tie the two running ends with a square knot and two overhand knots on the hip opposite the brake hand (Figure 1-3M, N).

(10) Place any excess rope in the trouser pocket near the square knot (Figure 1-3O, P).

(11) With the gate down and the hooked end of the snaplink against the navel, place the end of the snaplink through the single rope that is around the waist and the two ropes forming the double-hand knot (Figure 1-3Q).

(12) Rotate the snaplink a half turn so that the gate is facing up and will open away from the body (Figure 1-3R, S, T).

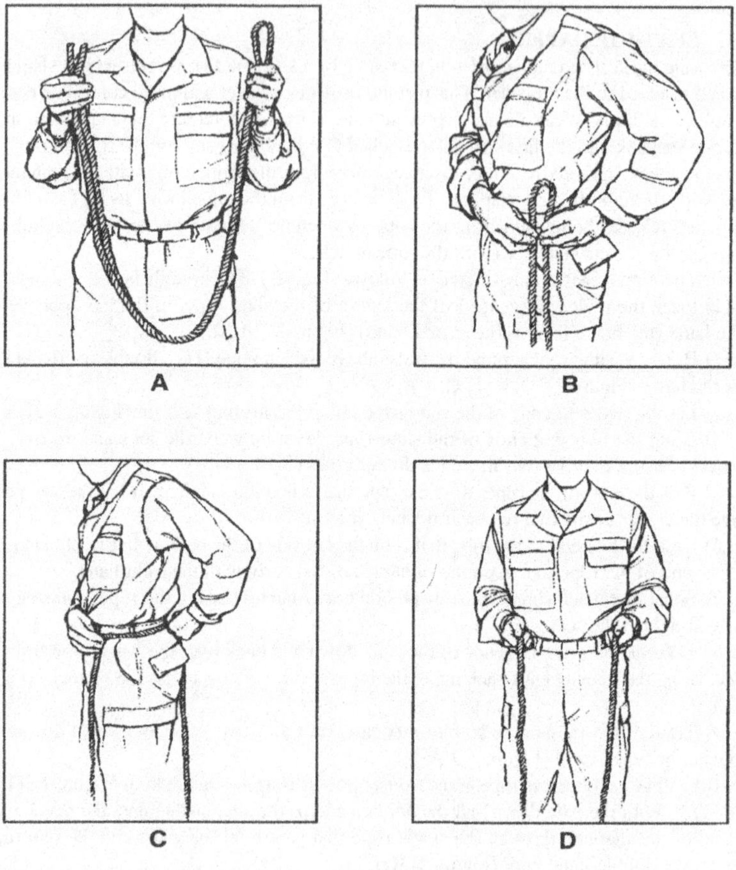

Figure 1-3. Construction of rappel seat.

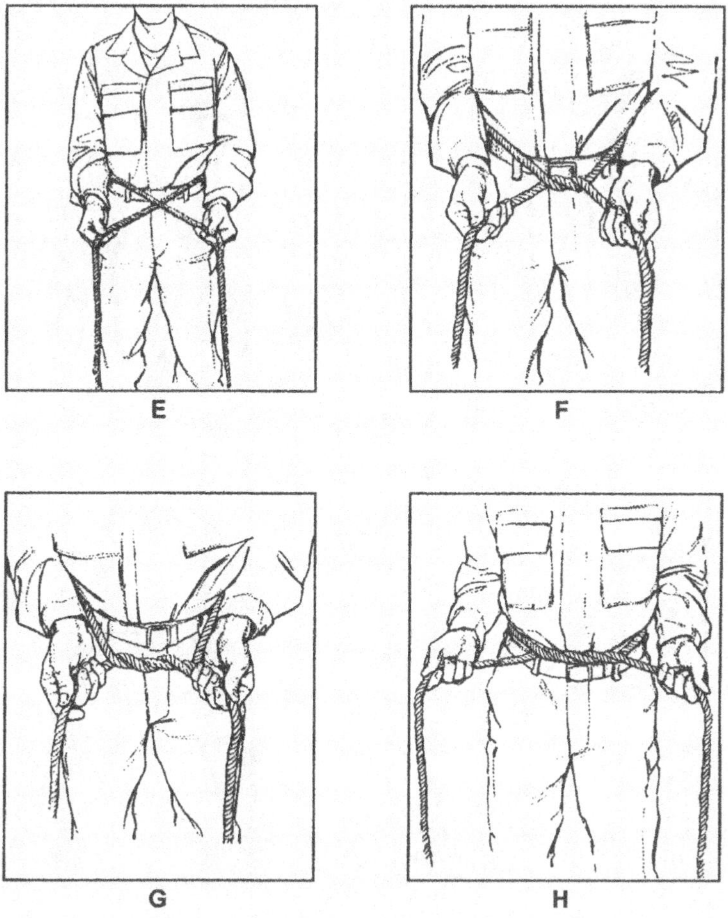

Figure 1-3. Construction of rappel seat.

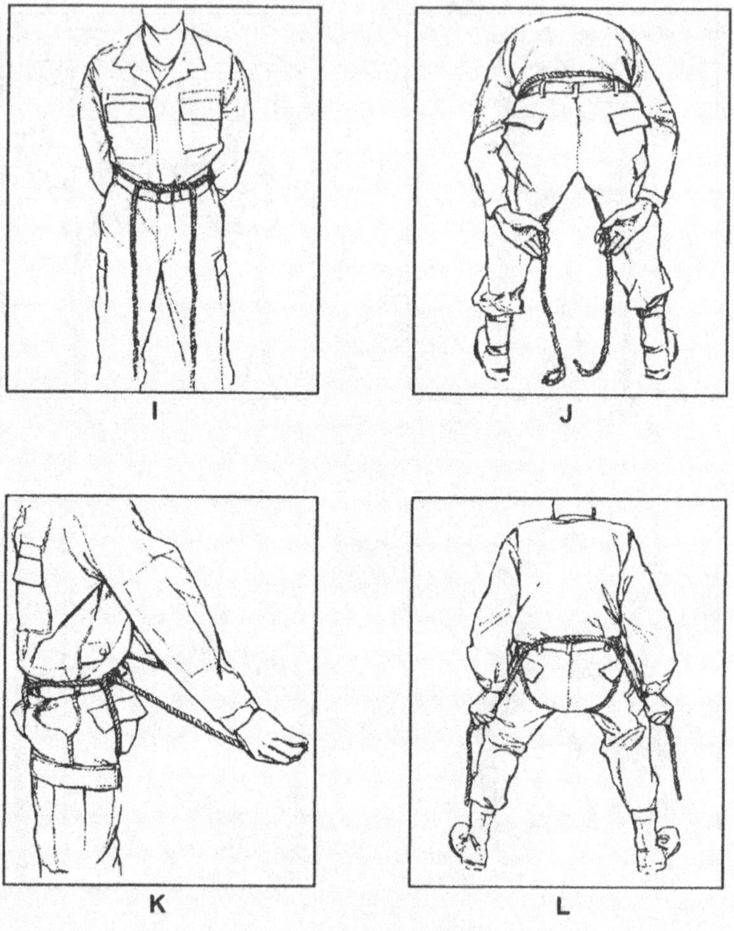

Figure 1-3. Construction of rappel seat.

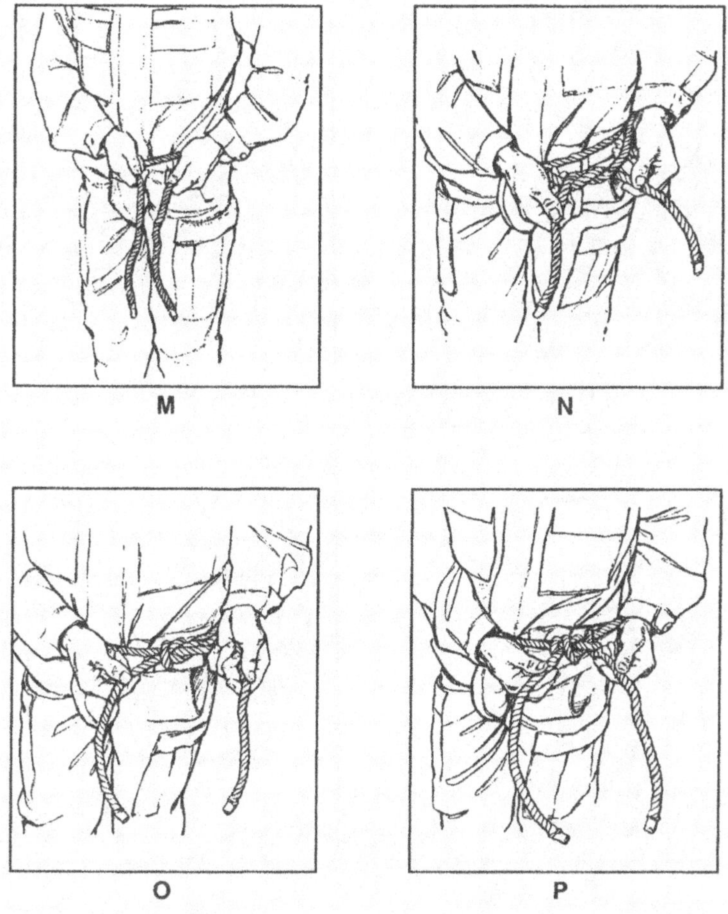

Figure 1-3. Construction of rappel seat.

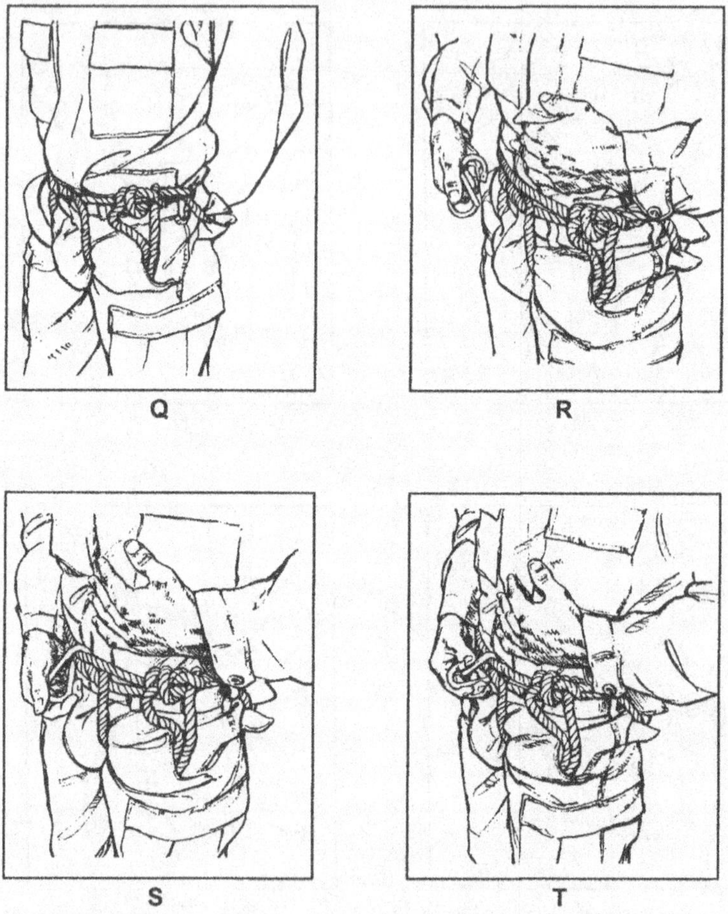

Figure 1-3. Construction of rappel seat.

b. To hook up using the seat-hip method, perform the following:

(1) Place the square knot with two overhand knots towards the anchor point for all seat-hip rappels.

(2) Grasp the two ropes with both hands and drop them through the gate of the snaplink. (At this point, two ropes should be running through the snaplink.)

(3) Using the hand closest to the anchor point, pull the slack towards the anchor point. Rotate the slack under and then over the top of the snaplink.

(4) Drop the two ropes a second time through the gate of the snaplink. (At this point, four ropes should be running through the snaplink.)

(5) Place the guide hand on the rope between the anchor point and the snaplink (palm facing up).

(6) Place the brake hand around the running end of the rope (palm facing down). Place the brake hand with the rope in the small of the back.

1-12. AUSTRALIAN RAPPEL

To hook up for the Australian rappel (Figure 1-4), perform the following:

a. Hook up to the Australian seat.

(1) Grasp the 9-foot sling rope at the midpoint (center) of its length, and double the rope.

(2) Place the doubled sling rope around the back and waist. Ensure that the rope is above the hipbone, but below the ribs.

(3) Tie a *square knot with two overhand knots* over the navel.

(4) Place any excess rope in the trouser pocket nearest the excess.

b. When the Australian seat is donned, face away from the anchor point and to the side of the rappel rope. (Stand to the same side of the rope as the brake hand. Determine left and right of the rope while facing the anchor point.) Place a snaplink onto the seat with the gate facing up, hinge closest to the body. Place it on the hip corresponding to the brake hand. The rappel master grasps the rappel rope and lays the rope into the snaplink. He then places one hand between the snaplink and the anchor point and draws slack toward the anchor point. He rotates the slack down, under, and over the rope and into the snaplink. The rappel master then slides the snaplink directly to the rappellers back.

c. Serve the running end of the rappel rope with the brake hand and prepare to rappel. During descent, brake by drawing the rappel ropes diagonally across the chest with the ropes running from near the waist to the pocket of the opposite shoulder.

Figure 1-4. Australian rappel.

1-13. CLIMBING PROCEDURES

Before climbing the ladder, the rappel master, safety OIC, or rappel lane NCO checks each rappeller's equipment.

a. The rappeller kicks the sand off his boots before climbing.

b. The rappeller grasps the outside of the ladder while climbing, not the rungs (when possible). If stairs are built for the tower, the rappeller grasps the railings as appropriate.

c. Just before climbing up the ladder, the rappeller sounds off, "(name) climbing," and then begins climbing up the ladder. Once at the top and clear of the ladder, the rappeller sounds off, "(name) clear."

d. Once off the ladder, the rappeller waits until the rappel master or lane NCO directs him to proceed to a rope station. At this time, the next rappeller in line may start to climb the ladder.

1-14. TOWER PROCEDURES

After the rappellers climb the tower, the following procedures are adhered to:

a. Once directed to a rope station, the rappel master or lane NCO ensures proper hookup for rappelling.

b. At this time, the rappeller sounds off with "on rappel" and the belayer sounds off with "on belay."

c. While maintaining his brake, the rappeller (on command from the rappel master or lane NCO) steps over the safety rail and faces the anchor point.

d. At this point, the rappel lane NCO sounds off with the following verbal commands and arm-and-hand signals.

(1) *Get Ready*—Rappel master or lane NCO extends both arms to the front with fists clenched and thumbs pointing upward (Figure 1-5). This alerts the rappeller. Each rappeller then looks over his brake hand shoulder to check for the belay man. The rappeller then looks at the rappel master. The rappel master or lane NCO makes his second check of the hookup, rappel seat, snaplink, and equipment.

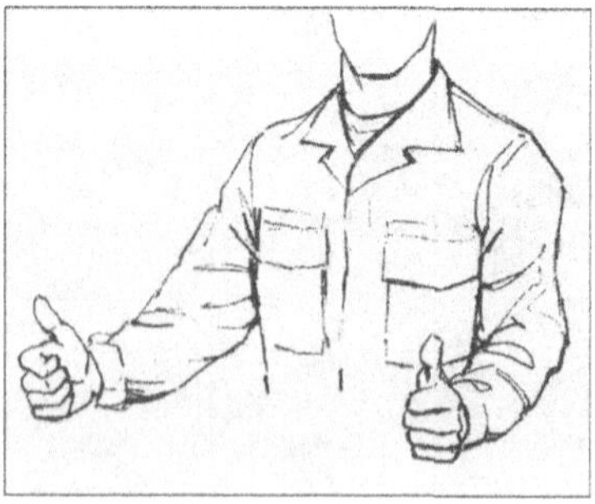

Figure 1-5. Example of arm-and-hand signal GET READY.

(2) *Position*—The lane NCO extends both arms to the front, elbows locked, forearms pointed downward, and fingers extended. He makes a circular motion with both forearms rotating in opposite directions (Figure 1-6). With the brake hand in the small of the back, the rappeller rotates 180 degrees out onto the wall or skid mock up and assumes an L-shaped position. The feet should be shoulder width apart, balls of feet on the wall or skid, knees locked, and body bent at the waist (Figure 1-7).

Figure 1-6. Example of arm and hand signal POSITION.

Figure 1-7. L-shape position.

(3) *Go*—The lane NCO extends his right arm with the elbow locked, fingers extended, thumbs around the index fingers, and points directly at the rappeller (Figure 1 8). This initiates the rappel. The rappeller flexes his knees and jumps vigorously backwards. At the same time, the rappeller throws his brake hand out at a 45-degree angle, letting the rope slide through both the brake hand and the guide hand. The rappeller looks over his brake hand shoulder at all times during descent.

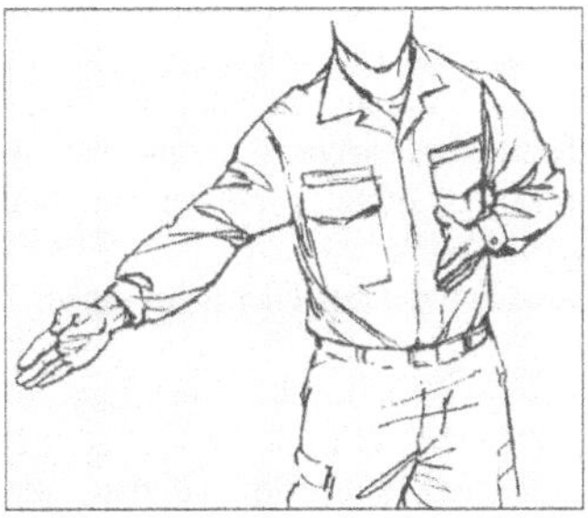

Figure 1-8. Example of arm-and-hand signal GO

e. The rappeller descends in a smooth, controlled manner.

f. The rappeller maintains eye contact with the ground at all times.

g. The rappeller maintains a modified L-shape position during descent with the feet shoulder-width apart, knees flexed, and buttocks parallel to the ground (Figure 1-9).

Figure 1-9. L-shape position while rappelling.

h. When carrying equipment or additional weight, a modified L-shape is used with the legs slightly lower than the buttocks to compensate for the added weight.

i. The rappeller's back is straight. He looks over the brake hand shoulder.

j. The guide hand is extended on the rope with the elbow extended and locked.

k. The rope slides freely through the guide hand, which is used to adjust equipment and to assist in balance during descent.

l. To brake, the rappeller places the brake hand (with rope in hand) in the small of the back and then grasps the rope firmly with the brake hand.

Note: Do not grip the rope firmly with the brake hand while the brake hand and brake arm are extended at the 45-degree angle. If this is done while rappelling, the brake hand and glove may become entangled in the snaplink causing injury to the hand and causing the rappeller to become hung up on the ropes.

m. Releasing tension on the rope and moving the brake hand out to the rear at a 45-degree angle regulates the rate of descent.

n. The rappeller never lets go of the rope with his brake hand until the rappel is completed.

o. After the rope is cleared and the rappeller is off rappel, he acts as the belayer for the next rappeller.

1-15. RAPPEL TOWER TRAINING FOR UH-1H HELICOPTER

Training on the rappel tower for helicopter skid rappelling prepares soldiers to rappel from a UH-1H helicopter (Figure 1-10). The rappeller is hooked up while he sits on the platform just above the helicopter skid. On the rappel master's command GET READY, the rappeller looks over the edge of the tower to ensure the running ends of the ropes are on the ground (Figure 1-10A). On the command SIT IN THE DOOR, the rappeller rotates his feet and legs off the platform and places them on the skid (Figure 1-10B). On the command POSITION, the rappeller turns around and assumes an L-shape position (Figure 1-10C). On the command GO, the rappeller bounds away from the helicopter skid and rappels to the ground. The rappel master is responsible for the proper procedures and safety.

A B C

Figure 1-10. UH-1H helicopter skid rappel training.

1-16. RAPPEL TOWER TRAINING FOR UH-60 BLACKHAWK HELICOPTER

Rappel tower training for the UH-60 Blackhawk is similar to that for the UH-1H with the exception that the UH-60 has no skids on which to stand. Therefore, the edge of the rappel tower is used as a pivot point to assume the L-shape position. All commands are the same except for SIT IN THE DOOR, which does not apply to the UH-60. The stances for each command are also different, as shown in Figure 1-11.

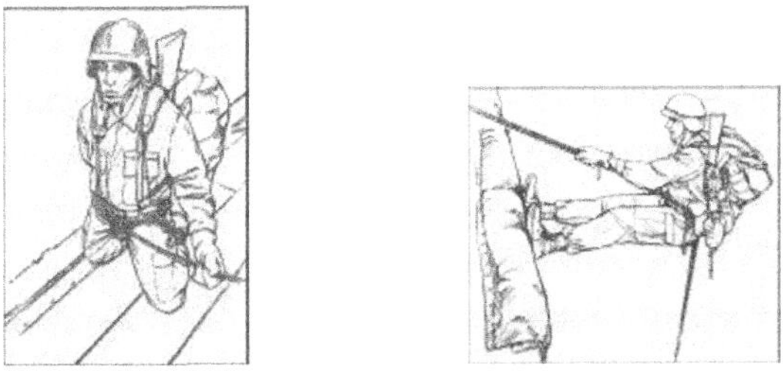

A B

Figure 1-11. UH-60 helicopter rappel training.

1-17. EMERGENCY LOCK-IN PROCEDURES

Mastering the lock-in procedure during tower training is critical before advancing to helicopter rappelling. Using the lock-in procedure allows personnel to hold in position for an extended period of time.

a. If the helicopter gains altitude above the length of the rappel ropes, the rappeller immediately brakes, locks in, and waits for the descent of the aircraft. Procedures for lock-in are as follows:

(1) Place the brake hand in the small of the back and brake to a complete stop.

(2) Release the guide hand from the ropes.

(3) Bring the guide hand around the back and grasp the running end of the two rappel ropes behind the brake hand.

(4) Using the guide hand, bring the two running ends of the rappel ropes around to the front.

(5) Secure these two running ends of the rappel ropes with the two anchor ends of the rappel ropes in the guide hand. This is now the new brake hand.

(6) Take the old brake hand out of the small of the back. Bring it around to the front and grasp the two ropes from the anchor point at a point just above the new brake hand. The old brake hand is now the new guide hand.

(7) Face the rappel master and wait for his command to lower to the ground.

(8) When the command is received from the rappel master to continue the descent, bring the brake hand to a 45-degree angle to the rear. When it is time to brake, bring the new brake hand around to the front diagonally across the chest.

b. If an engine fails or an aircraft emergency occurs during rappelling, the rappellers on the ropes descend as rapidly as possible and move from beneath the aircraft to the sides. Rappellers will maintain control of ropes, if possible.

1-18. COMMUNICATIONS

The rappeller at the top of a rappel point must be able to communicate with those at the bottom. Radios, hand signals, and rope signals are considered during a tactical rappel. For training situations, use the following commands:

COMMAND	GIVEN BY	MEANING
LANE NUMBER, ON RAPPEL	Rappeller	I am ready to begin rappelling.
LANE NUMBER, ON BELAY	Belayer	I am on belay and you may begin your rappel.
LANE NUMBER, OFF RAPPEL	Rappeller	I have completed the rappel, cleared the rappel lane, and am off the rope.
LANE NUMBER, OFF BELAY	Belayer	I am off belay.
LANE NUMBER, FALLING	Rappeller	I am falling. Be alert below—belay man brake.

1-18. DEMONSTRATION

After explaining the procedures to all rappellers, the rappel master should have an assistant demonstrate one complete cycle of rappelling from the static tower. This ensures that the rappellers can hear all the proper commands and see the actions and techniques used on the static tower.

CHAPTER 2
GROUND RAPPELLING

Ground rappelling is a technique that allows soldiers to negotiate mountains and cliffs safely and rapidly. Before rappellers participate in mountain and cliff rappelling they should have completed all the requirements for tower training. To ensure the rappel is safely conducted and meets the mission requirements, rappellers should consider several factors.

2-1. PERSONNEL

The personnel responsibilities for mountain and cliff rappelling are the same as for tower rappelling. As a minimum the following personnel are needed for mountain and cliff rappelling.

a. **Mountaineering Safety Officer (MSO).** The MSO is a SFC or above who is ranger qualified or is a graduate of the Northern Warfare Training Center (NWTC). He has overall responsibility for the safety of all mountaineering participants and ensures that all safety precautions are followed.

b. **Rappel Master**. The rappel master is a qualified officer or NCO in charge of training on ground, tower, and aircraft rappel safety procedures.

c. **Medical Officer or Noncommissioned Officer.** Qualified rappellers and aidmen must know MEDEVAC procedures.

d. **Radiotelephone Operator.** The RATELO monitors the radio and reports to the RSO.

e. **Rappel Lane NCO.** The rappel lane NCO is a qualified officer or NCO in charge of rappelling on a wall or ground rappel lane (Chapter 1, paragraph 1-3).

2-2. SUSTAINMENT TRAINING

Before conducting ground rappel training, the unit or element conducts sustainment training.

a. Review the construction of a rappel seat, seats to be used, and hook-up procedures.

b. Conduct two rappels on the 34-foot tower wall (20-foot tower is acceptable): one with equipment, one without equipment. Conduct two rappels from the free side of a 34-foot or higher rappel tower.

c. Demonstrate the ability to lock in during one of the rappels from the free side of the tower.

2-3. SELECTION OF A RAPPEL POINT

The selection of the rappel point depends on factors such as mission, cover, route, anchor points, and edge composition (loose or jagged rocks). The anchor point should be above the rappeller's departure point. Primary and secondary anchors must be solid (natural anchors are preferred). The rappeller should be sure that the rope reaches the bottom or a place from which he can further rappel or climb. Also, the rappel point should be carefully tested and inspected to ensure the rope will run freely and that the area is clear of obstacles that could be dislodged. If a sling or runner is used for a rappel point, it should be tied twice from two separate loops. Suitable loading and off-loading platforms should be available.

2-4. ESTABLISHMENT OF A RAPPEL POINT

A rappel lane is the area and rope that a rappeller will use during his descent. It should have equal tension between all anchor points by establishing primary and secondary anchor points (Figure 2-1). If one anchor point fails, the rappel rope should not extend. All of the methods discussed below can be performed with a single or double rope. A double rope application should be used for safety.

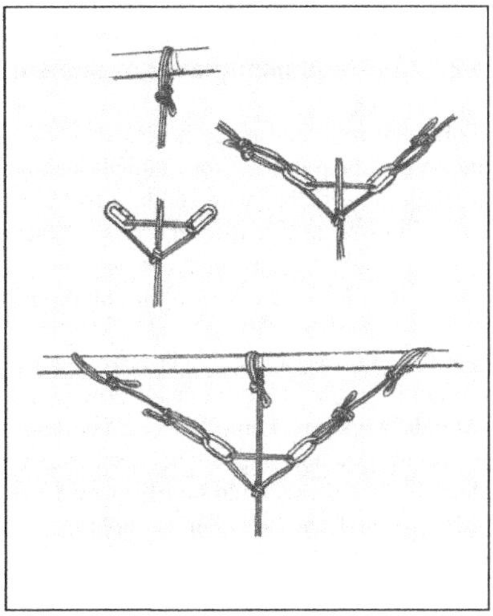

Figure 2-1. Rappel point.

a. If a rappel lane is *less* than half the rope length, the rappeller may apply one of the following techniques:

(1) Double the rope and tie a three-loop bowline around the primary anchor to include the primary anchor inside two loops and enough rope in the third loop to run to the secondary anchor (another three-loop bowline secured with an overhand knot).

(2) Double the rope and tie it around a secure anchor point with a round turn anchor bowline secured with an overhand knot (or any appropriate anchor knot).

(3) Double the rope and establish a self-equalizing anchor system (Figure 2-2) with a bowline-on-a-bight or figure-eight-on-a-bight knot. Tie off on the long standing and with a round turn anchor bowline.

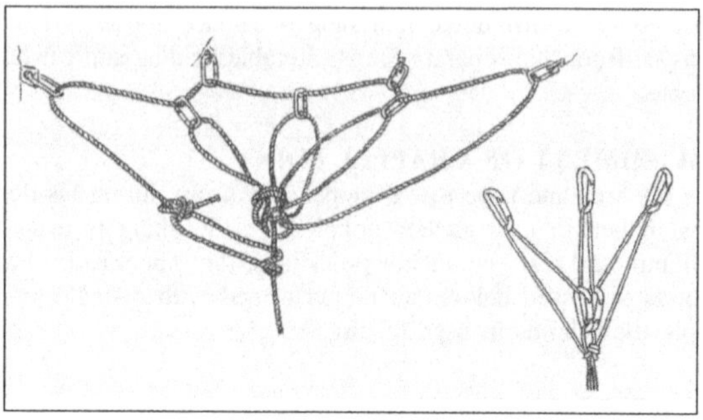

Figure 2-2. Self-equalizing anchor system.

(4) In an emergency, double the rope and place it behind or through a secure anchor point, or tie a runner around an anchor point with a snaplink inserted and place the rope through the snaplink. To preclude a rappeller from sliding off the end of the rappel lane, tie a double figure eight (square knot or double fisherman's knot) at the bottom end of the rope with both ends.

b. If a rappel lane is *greater* than half the rope length, the rappeller may apply one of the following techniques:

(1) Using two ropes, tie a round turn anchor bowline around a primary anchor point. Take the remaining rope (the tail from the primary anchor bowline) and tie another round turn anchor bowline to a secondary anchor point. The secondary anchor point should be in a direct line behind the primary anchor point. The anchor can be either natural or artificial. The ends of the rappel lane ropes should be offset by 15 centimeters (6 inches) so that the rope ends feed freely through the rappeller's snaplink.

(2) Using two ropes, establish a three-piton anchor system using a bowline-on-a-bight knot (or figure-eight-on-a-bight knot) and tie off on the long-standing end with a round turn anchor bowline.

(3) In an emergency, use two ropes and tie the two ends together with a joining knot. Place the joined ropes behind or through an anchor point, or tie a runner around an anchor point with a snaplink inserted and place the joined rope through the snaplink. Offset the joining knot to the left or right of the anchor. Tie off the bottom end of the rope with a joining knot to prevent a rappeller from sliding off the end of the rappel line.

c. Due to the length of the rappel, the rappel rope may not reach the anchor. If the rope is used to tie the knots, it may be too short to accomplish the rappel.

(1) When using a natural anchor—

(a) Tie a sling rope, runner, or another rope around the anchor with a round turn anchor bowline.

(b) Tie a fixed loop (figure eight or butterfly) in one end of the rappel rope, which is attached to the round turn around the anchor through the two snaplinks (opposing gates).

(2) When using an artificial anchor—

(a) Tie off a sling rope, runner, or another rope to form a loop.

(b) Put the loop through the snaplinks that are attached to the artificial anchor point.

(c) Bring the bottom of the loop up and connect it to the snaplinks that are between the artificial anchors (chocks, pitons, or bolts).

(d) Grasp the snaplinks that are between the chocks/pitons and pull them down and together.

(e) Tie a fixed loop (figure eight or butterfly) in the end of the rappel rope and connect this to the snaplinks that have been pulled together.

Notes: 1. Rerouted figure-eight knots can be used instead of bowlines.
2. Runners may be used from one or more anchor points.

d. Backfeed (stack) the rope to ensure it does not snarl when thrown. Take off one wrap at a time and let it fall to the ground, ensuring that no kinks, knots, or twists occur that might hinder the rope from feeding out. When the rope is backfed, anchor off one end of the rope. The two methods used to throw the rope are underhand and overhand. Use the overhand method when trees or shrubs are on or near the rappel point.

e. Set up a retrievable rappel point by applying one of the following techniques:

(1) When the rappel is *less* than half the total length of the rope, double the rope. Place the rope around the primary anchor with the bight formed by the midpoint. Join the tails of the rappel rope and throw the rope over the cliff. Tie a clove hitch knot around a snaplink just below the anchor point ensuring the locking bar inside the snaplink points away from the gate opening end and faces uphill. Snap the opposite standing portion into the snaplink. Upon reaching the bottom of the cliff, pull on the part of the rope to which the snaplink is secured to allow the rope to slide around the anchor point.

(2) When the length of the rappel is *greater* than half the length of the rope used, join two ropes around the anchor point (double fisherman's knot or square knot). Adjust the joining knot so that it is away from the anchor. Tie a clove hitch knot around a snaplink just below the anchor point ensuring the locking bar inside the snaplink points away from the gate opening end and faces uphill. Snap the opposite standing portion into the snaplink. Upon completion of the rappel, pull the rope to which the snaplink is secured to allow the rope to slide around the anchor point.

WARNING

When setting up a retrievable rappel using only a primary point, take care in selecting the point. Ensure soldiers have a safety line when approaching the rappel point, with only the rappeller going near the edge.

2-5. TYPES OF RAPPELS

Types of rappels include body rappel, hasty rappel, seat-hip rappel, seat-shoulder rappel, Australian rappel, and buddy-evacuation. (Seat-hip rappel and Australian rappel are described in Chapter 1, paragraphs 1-10 and 1-11.)

a. **Body Rappel.** Face the anchor point and straddle the rope. Pull the rope from behind and run it around either hip, diagonally across the chest, and back over the opposite shoulder (Figure 2-3). Then, run the rope to the brake hand, which is on the same side of the hip that the rope crosses (for example, the right hip to the left shoulder to the right hand). Lead with the brake hand down and face slightly sideways. The foot corresponding to the brake hand precedes the guide hand at all times. Keep the guide hand on the rope above to guide—not to brake. Lean out at a sharp angle to the rock. Keep the legs spread well apart and relatively straight for lateral stability and the back straight to reduce friction. Turn the BDU collar up to prevent rope burns on the neck. Wear gloves, and use other clothing to pad the shoulders and buttocks. To brake, lean back and face directly toward the rock area so the feet are horizontal to the ground.

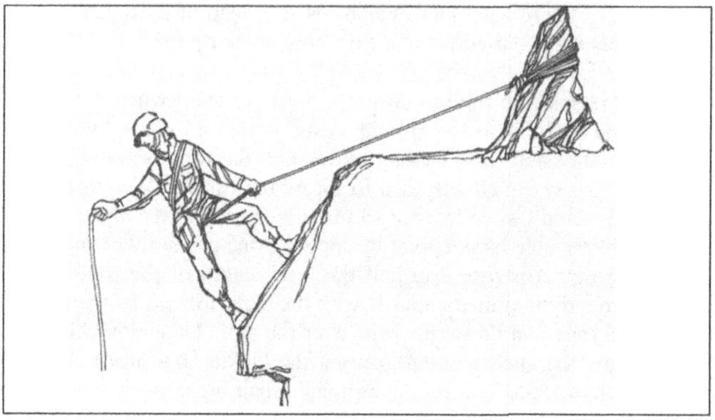

Figure 2-3. Body rappel.

b. **Hasty Rappel.** Facing slightly sideways to the anchor, place the ropes horizontally across the back (Figure 2-4). The hand nearest to the anchor is the guide hand, and the other is the brake hand. Wear gloves to prevent rope burns. To stop, bring the brake hand across in front of the body locking the rope. At the same time, turn to face up toward the anchor point. Use this rappel only on moderate rock pitches. The hasty rappel's main advantage is that it is easier and faster than the other methods, especially when the rope is wet.

Figure 2-4. Hasty rappel.

Figure 2-5. Seat-shoulder rappel.

c. **Seat-Shoulder Rappel.** To hook up for the seat-shoulder method, face the rappel point (Figure 2-5). Snap into the rope that passes up through the snaplink. Bring the rope over one shoulder and back to the opposite hand (left shoulder to right hand). Use the same technique in the descent as in the body rappel. This method is faster than the body rappel, less frictional, and more efficient rappellers with packs and during night operations.

d. **Buddy-Evacuation Rappel.** Use the buddy-evacuation rappel to evacuate an injured soldier from a cliff or steep terrain. Face the cliff and assume a modified L-shape body position to compensate for the weight of the victim on the back. The victim is top-rope belayed from above, which provides the victim with a point of attachment to a secured rope.

Note: To use this rappel, the victim must be conscious.

(1) The method for securing a victim to a rappeller's back is described below:

(a) To secure the victim to the carrier's back with a rope, the carrier ties a standard rappel seat (brake of choice, depending on the injury) and rests his hands on his knees while the victim straddles his back (Figure 2-6.)

(b) A 4.2-meter (14-foot) sling rope is used. A 45-cm (18-inch) tail of the sling is placed on the victim's left hip. (This method describes the procedure for a seat-hip rappel with right-hand brake.)

(c) The long remaining end of the sling rope is routed under the victim's buttocks and passed over the victim's and carrier's right hip. The rope is run diagonally, from right to left, across the carrier's chest, over his left shoulder, and back under the victim's left armpit.

(d) The rope is then run horizontally, from left to right, across the victim's back. The rope is passed under the victim's right armpit and over the carrier's right shoulder.

Figure 2-6. Patient secured to carrier's back.

(e) The rope is run diagonally, from right to left, across the carrier's chest and back across the carrier's and victim's left hip.

(f) The two rope ends should now meet. The two ends are tied together with a square knot and overhand knots.

(g) The knot is positioned on the victim's left hip. The carrier's shoulders should be padded to prevent cutting by the rope.

(2) An alternate method is to use two pistol belts hooked together and draped over the carrier's shoulders. The victim straddles the carrier, and the belay man secures the loose ends of the pistol belts under the victim's buttocks. Slack in the pistol belt's sling should be avoided, since the carrier is most comfortable when the victim rests high on his back. (See FM 8-35.)

(3) A large rucksack can be slit on the sides near the bottom so that the victim can step into it. The victim is belayed from the top with the carrier conducting a standard rappel. The carrier wears the rucksack with the victim inside.

Note: Rucksacks are not designed to support the weight of a soldier and his gear. Therefore, this technique is used only as a last resort.

(4) A casualty secured to a carrier, as described above, can be rappelled down a steep cliff using a seat-shoulder or seat-hip rappel. The casualty's and rappeller's shoulders should be padded where the sling rope and rappel lines cross if a seat-shoulder rappel is used. The buddy team should be belayed from above with a bowline tied around the victim's chest under his armpits. The belay rope must run over the rappeller's guide hand shoulder (Figure 2-7).

Figure 2-7. One-man carry.

2-6. RAPPELLING PROCEDURES

Proper rappelling procedures must be followed to ensure the safety of personnel.

a. The rappel lane NCO is tied off to a separate anchor point for safety, if possible. He—

(1) Ensures that the anchors are sound and the knots are properly tied.

(2) Ensures that loose rock and debris are cleared from the loading platform.

(3) Allows only one man on the loading platform at a time and ensures that the rappel point is orderly run.

(4) Ensures that each man is properly prepared for the particular rapel: gloves on sleeves down, helmet with chin strap fastened, gear prepared properly, and rappe seat and knots correct (if required). He also ensures that the rappeller is correctly hooked up to the rope and is aware of the proper braking position.

(5) Ensures that the proper signals or commands are used.

(6) Dispatches each man down the rope.

(7) Is the last man down the rope.

(8) Ensures that the rope does not run over sharp edges, and pads the rappelling surfaces as necessary to protect the rope.

(9) Ensures that the rope reaches the bottom or is at a place where additional rappels can be made.

b. The first rappeller down—

(1) Selects a smooth route for the rope that is clear of sharp rocks.

(2) Conducts a self-belay with a prusik sling tied from himself to the rappel rope.

(3) Clears the route and places loose rocks, which the rope may dislodge, far enough back on ledges to be out of the way.

(4) Ensures the rope reaches the bottom or is at a place from which additional rappels can be made.

(5) Ensures that the rope runs freely around the rappel point when pulled from below.

(6) Clears the rappel lane by straightening all twists and tangles from the ropes.

(7) Belays subsequent rappellers down the rope.

(8) Takes charge of personnel as they arrive at the bottom (off-loading platform).

Note: A rappeller is always belayed from the bottom except for the first man down. The first man belays himself down the rope by using a safety line attached to his rappel seat that is hooked to the rappel rope with a prusik knot. As the first man rappels down the rope, he "walks" the prusik knot down with him.

2-7. DUTIES OF THE RAPPELLER

To ensure the safe and efficient conduct of rappelling operations, each rappeller must know the duties of his job.

a. Each rappeller down shouts, "Off rappel" (if the tactical situation permits), untangles the ropes, and ensures the ropes run freely around their anchors. When silence is needed, a planned signal of pulling the rope is substituted for the voice signal. After the rope is cleared and the rappeller is off rappel, he acts as the belayer for the next rappeller.

b. All rappellers inspect ropes often when many soldiers are rappelling.

c. The last rappeller to descend constructs a retrievable rappel point and rappels down. Then he gently pulls the rope to prevent the rising rope end from entangling with the other rope. He stands clear of the falling rope and any rocks that it may dislodge.

Note: Soldiers wear gloves for all types of rappels for protection from rope burns. Also, bounding rappels are discouraged since this stresses the anchor and causes undue wear and friction on the rope.

d. Rappellers descend in a smooth, controlled manner. The body forms an L-shape with the feet shoulder-width apart, legs straight, and buttocks parallel to the ground. When carrying equipment or additional weight, a modified L-shape is used with the legs slightly lower than the buttocks to compensate for the added weight. The rappeller's back is straight. He looks over the brake shoulder. The guide hand is extended on the rope with the elbow extended and locked. The rope slides freely through the guide hand, which is used to adjust equipment and to assist balance during the descent. The rappeller grasps the rope firmly with the brake hand and places it in the small of his back. Releasing tension on the rope and moving the brake hand out to his rear at a 45-degree angle regulates the rate of descent. The rappeller never lets go of the rope with his brake hand until the rappel is completed.

e. During training, the lane number must be understood. In combat, a series of tugs on the rope may be substituted for the oral commands to maintain noise discipline. The number of tugs used to indicate each of the commands is IAW the unit SOP.

2-8. BELAYER

The belayer assumes a position at the base of the lane about one pace away from the rock area. He ensures that the rappel ropes are even with the ground during rock and tower rappels. The belayer loosely holds the rappel ropes with both hands so as not to interfere with the rappeller but still able to stop the rappeller should he fall. If the rappeller shouts "Falling" or loses control of his brake hand or descent, the belayer immediately stops the rappeller by pulling downward on the rappel ropes. To ensure a firm grip on the rappelling rope, the belayer does not wear gloves. (Because no friction exists between the belayer's hands and the rappelling rope, gloves are not required for safety. However, whether or not to wear gloves is decided by the OIC/NCO.) The belayer watches the rappeller at all times and maintains constant voice or visual contact with the rappeller. The belayer wears a helmet to prevent injuries from falling debris. All commands are spoken loudly and clearly.

CHAPTER 3

HELICOPTER RAPPELLING

Helicopter rappelling can provide a means of quick insertion with or without an LZ.

Section I. PERSONNEL

The personnel required for helicopter rappelling are the rappel master, the rappel safety officer, the pilot in command, the rappelers, and the belayers.

3-1. RAPPEL MASTER

A qualified rappel master is aboard each aircraft. Safety is the rappel masters number one priority. The rappel master—

- Ensures internal communication between the pilot and rappel master, and external communication between the aircraft and the ground.
- Inspects all equipment and uses only authorized, serviceable equipment.
- Inspects and tests all anchor points and knots before the mission starts.
- Ensures that all rappellers receive a safety briefing and the pilots and aircrew receive an air mission brief.
- Ensures that the rappelers are rappel qualified before conducting helicopter rappelling to include tactical rappelling.
- Maintains communications with the pilot at all times.

3-2. RAPPEL SAFETY OFFICER

The RSO is a SFC or above who is either air assault or ranger qualified. He has overall responsibility for the safety of all rappellers and ensures that all safety precautions are followed. He maintains communications at all times with the pilot and rappel master through FM radio. He alerts the rappel master and pilot of any unsafe acts.

3-3. PILOT IN COMMAND

The pilot in command (PIC) of the aircraft has the following responsibilities:

a. Ensures that the aircrew and all non-aircrew personnel are briefed and understand their responsibilities during rappelling operations, including aircraft safety and action in the event of an emergency.

b. Ensures that the donut ring anchoring device assembly and or aircraft anchor points have been inspected for completeness and functionality, and that they are installed properly.

c. Emphasizes procedural techniques for clearing, recovery, and or jettison of ropes.

d. Keeps the aircraft centered over the target with corrections from the rappel master as required.

3-4. RAPPELLER

In addition to the tower qualification requirements outlined in Chapter 1, the individual rappeller must complete advance training under the supervision of a qualified rappel master in order to participate in tactical helicopter rappelling operations. He must—

- Satisfactorily complete three rappels from a helicopter from a height of 60 feet (two rappels with combat equipment and weapon).
- Demonstrate confidence and proficiency in the techniques, procedures, and equipment used in rappelling from a helicopter.
- Know the rappelling equipment used in helicopter operations and any special equipment required for helicopter rappelling.

3-5. BELAYER

A belayer is assigned to each rope. He is responsible for walking the rope beneath the helicopter during the descent. (Walking the rope is defined as removing the slack from underneath the helicopter by walking backwards with the rope as the helicopter descends to land.) The belayer makes sure the ropes are not caught on the aircraft skids or tires to ensure a safe landing.

Section II. TRAINING

Training for helicopter rappelling includes sustainment training and refresher training.

3-6. SUSTAINMENT TRAINING

Before conducting helicopter rappel training, the unit or element conducts sustainment training.

a. Review the construction of a rappel seat, equipment to be used, and hook-up procedures.

b. Conduct two rappels on the 34-foot or higher tower wall: one without equipment, one with equipment. Conduct two rappels from the open side of a 34-foot or higher rappel tower.

3-7. REFRESHER TRAINING

Refresher training is routinely conducted to maintain acquired skills. Soldiers who have not performed a helicopter rappel during the past six months will undergo refresher training consisting of three satisfactory rappels from a tower (one with weapon and equipment and one executing a lock in) before executing a helicopter rappel.

Section III. PREOPERATIONAL BRIEFINGS AND SAFETY PROCEDURES

This section discusses safety, medical and communications requirements, and the procedures to follow during unusual conditions (adverse weather/terrain conditions, night operations). Personnel must use sound judgement to determine what action to take depending on the nature and severity of the conditions.

3-8. MEDICAL COVERAGE

A qualified and equipped medic will be present to respond to any mishap. Medical transportation must also be available. Absence of a medic, medical equipment, or transportation will terminate training. If the situation warrants and the installation cannot support a MEDEVAC mission, the rappel aircraft may be used as a last resort MEDEVAC vehicle.

3-9. COMMUNICATION REQUIREMENTS

During helicopter rappel training, the RSO will have radio communications with the aircraft. Voice communications are required before starting aircraft rappelling. Additionally, the RSO will inform the PIC to stop operations if an unsafe condition develops.

3-10. ADVERSE WEATHER/TERRAIN CONDITIONS

Rappel operations will not be conducted under the following conditions

- Ambient temperature is 30 degrees Fahrenheit or less.
- Winds in excess of 30 knots.
- Lightning strikes within 1 nautical mile of rappelling operations.
- Wind chill factors caused by the helicopter's rotor wash or extraction cruise air speeds, which could cause cold weather injuries.
- Water or ice on the rope inhibiting the ability of the rappellers to control their descent.
- The rope is exposed to the elements for a sufficient length of time to freeze, thereby reducing its tensile strength.
- Blowing particles produced by rotor wash causes the aircrew or the rappel master to lose visual contact with the ground.

3-11. NIGHT OPERATION REQUIREMENTS

The following requirements are necessary for night rappelling operations.

a. One chemlight will be attached to the end of the rope and one on each rappeller.

b. One chemlight will be secured to the attachment point of the rope.

c. Night vision goggles (NVG) will not be worn by rappellers during the descent. Aircrew members will wear NVG as required during night operations. NVG lighting criteria will be IAW Army regulations, specific aircraft aircrew training manuals, unit SOPs, or the tactical environment.

3-12. SAFETY BRIEFING

The following safety measures are enforced.

a. Loose clothing and equipment are secured.

b. Helmets are worn with chin straps fastened.

c. Rappellers wear identification tags and earplugs, carry identification cards, and role down their sleeves.

d. Weapons are slung diagonally across the back with the muzzle pointing down on the guide hand side.

e. All seats and rappelling equipment must be inspected by a rappel master before rappelling.

f. No running is allowed within 50 feet of the aircraft.

g. Personnel approach and depart the helicopter from the front and forward of the rear of the cargo doors. When approaching or departing the helicopter, personnel bend their bodies forward at the waist to ensure clearance of the rotor blades. At no time will personnel go near the rear of the aircraft.

h. Upon boarding the aircraft, the rappeller sits or kneels down, hooks up, and applies his brake hand to the small of his back.

i. While in the helicopter, the rappeller maintains eye-to-eye contact with the rappel master and receives commands from him.

j. The rappeller ensures that he has a belayer on his rope at all times when conducting training at a hover site.

k. During the descent, the rappeller maintains eye-to-ground contact.

l. If the rappeller sees his rope coming off the ground or sees that his belayer has lost control of his rope, he immediately brakes and executes a lock-in. He then waits for commands from the rappel master.

m. The rappeller brakes once every 30 feet during descent.

n. The belayer does not wear gloves. He keeps both hands on the rope and his eyes on the rappeller at all times.

o. All rappelling will be conducted using a double strand of rope.

Section IV. DEPLOYMENT OF ROPES

Deployment of the ropes from a helicopter is a critical task. It can cause a planned rappelling operation to fail, or it can increase the time required to conduct the operation. This is due to the likelihood of the ropes becoming entangled. To prevent this, ropes must be deployed using a positive control technique. Two of the techniques that may be used are the *deployment bag* technique and the *log coil* technique.

3-13. DEPLOYMENT BAG TECHNIQUE

The rappeller places the deployment bag (D-bag) (standard) on a flat surface with the stow loop facing upward. If the D-bag still has a static line, the rappeller removes it by cutting the static line where it attaches to the bag. A 12-foot section of the static line can be used as a safety line in the helicopter.

a. Use a sandbag (about one E-tool spade full of sand) as a weight in the deployment system. Roll the sandbag into a rectangular shape, tie it, and then place it in a small plastic bag, rolled and taped. Place a retaining band over the middle of the weight, and place the weight on top of the D-bag. The weight should be about the same width as the D-bag.

b. Lay out and inspect the two ropes. The working ends (closest to the D-bag) should be even. Place a round turn with the two working ends of the ropes on the weight's retaining band. Working on top of the deployment bag, start forming figure eights. The stack should consist of 8 to 10 figure eights, one on top of another—do not exceed the width of the D-bag. Then, starting from either side, center a retaining band over the stack. Ensure that it is over all the figure eights in the stack. Repeat the process each time, placing one stack in front of the other. Continue until about 10 feet of rope remains. If one rope is shorter than the other, the end of the shorter rope should be about 10 feet from the last stack.

c. For a primary anchor point, measure down about 4 feet from the end of the shortest rope. Using both ropes, tie a bowline without a half hitch. This knot is the primary anchor point.

d. For a secondary anchor point, tie an end-of-line bowline with a half hitch toward the end of the ropes. The dressed knot should be 18 to 22 inches from the primary anchor point knot when using the UH-1H helicopter and 22 to 30 inches when using the UH-60 helicopter. If excess rope remains, S-fold and tape it between the two knots. Ensure that 3 to 6 feet of rope remain from the last stack of figure eights.

e. To place the ropes in the D-bag, remove the bag from under the stack of rope. Place the weight into the bottom of the D-bag, and place all the stacks of figure eights, in the order they were made, one on top of the other into the bag.

f. Two left, two right, and two rear loops are in the opening of the D-bag. ***Do not use the two rear loops***. Close the flap on the D-bag and push the loops through the aligning holes on the flap. Using an 8-inch piece of gutted 550-pound cord, tie the two loops together with a square knot and two hitches. Ensure that two 120-foot ropes are coming out of the center of the flap—not to one side. Repeat the process and tie the right loops together. Wrap the excess rope lengthwise around the bag.

Note: The sizes of the loops of the bowline should be no larger than an average size fist.

3-14. LOG COIL TECHNIQUE

The rappeller lays the running end of the double rope along the length of the coiling log (Figure 3-1). He then coils the double rope around both the running end of the rope and the coiling log. The rope must be coiled evenly and tightly.

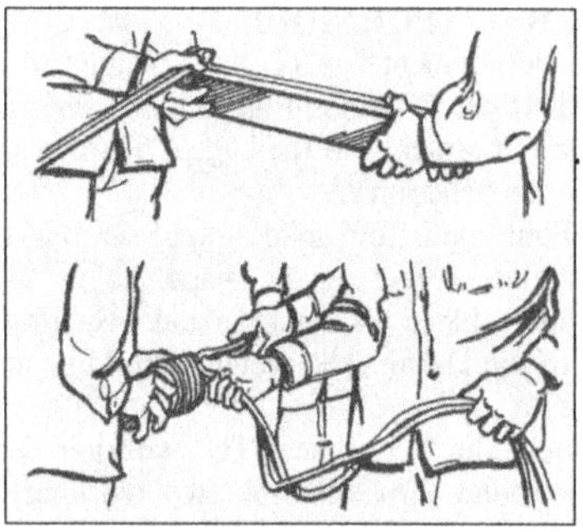

Figure 3-1. Coiling Log

Section V. RAPPELLING OPERATIONS FOR UH-1H IROQUOIS HELICOPTER

The UH-1H helicopter provides a safe, stable aerial platform from which to conduct rappelling operations when landing is not feasible.

3-15. CHARACTERISTICS

The UH-1H is a single-engine, medium-speed, single-main rotor helicopter that can transport eight rappellers, one rappel master, and one three-man crew. It has a lift capacity of 2,300 pounds and is equipped with several floor-mounted tie-down fittings, seven of which are used during rappelling operations.

3-16. RIGGING OF UH-1H HELICOPTER FOR RAPPELLING

The rappel master performs the following to rig the UH-1H helicopter.

a. Removes all seats.

b. Locks the doors in the open position. If no locks are present, removes doors. Also, he removes small cargo doors.

c. Pads and tapes all sharp edges on the floor, door ledge, and all protrusions on the skids. Ensures each door ledge has a scuff pad to protect the rope from contacting the metal door ledge.

d. Secures the donut ring to the center of the floor of the helicopter. The donut ring has six snaphooks numbered clockwise, with 12 o'clock being toward the front of the helicopter. Positions the clamps of the donut ring toward the aft end of the helicopter. Ensures the front two snaplinks and rear two snaplinks are facing outside of the donut ring and hooks them to the tie-down ring with the snaphook facing down. Ensures the center two snaphooks face into the donut ring and are connected to the floor tie-down ring with the gate facing down (Figure 3-2). Attaches the free-floating safety ring to the center floor tie-down ring using a seventh snaphook.

Note: Using this technique has caused equipment failure in many aircraft. The cable clamp nuts must be inspected and certified as airworthy by the pilot or maintenance crew before flying the mission.

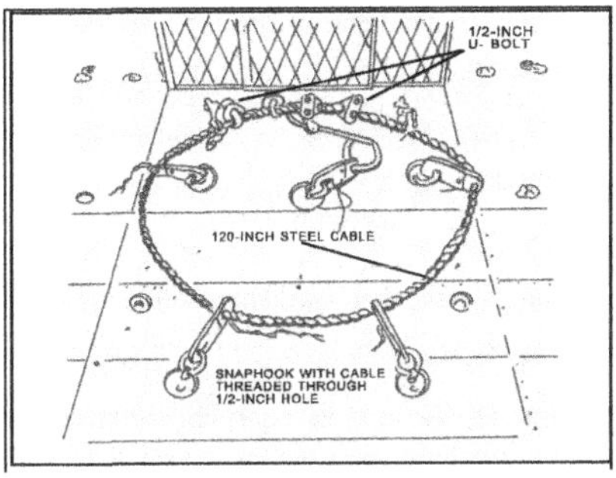

Figure 3-2. Donut ring attached to the floor of the helicopter.

3-17. CONSTRUCTION OF ANCHOR POINTS

In addition to the usual equipment requirements for rappelling operations, an anchor assembly is fabricated. This anchor assembly is commonly known as a *donut ring*. The secondary anchor point is a *floating safety ring*.

a. **Donut Ring.** This primary (or No. 1) anchor point for the rappelling ropes is constructed from a 1/2-inch steel cable with a steel wire core. The cable is 120 inches long, consists of 6 strands (18 wires per strand), and has a tensile strength of 21,000 pounds (Figure 3-2). The completed donut ring has a tensile strength of 3,000 pounds.

(Figure 3-3 shows the donut ring and rappel rope connection in a helicopter.)

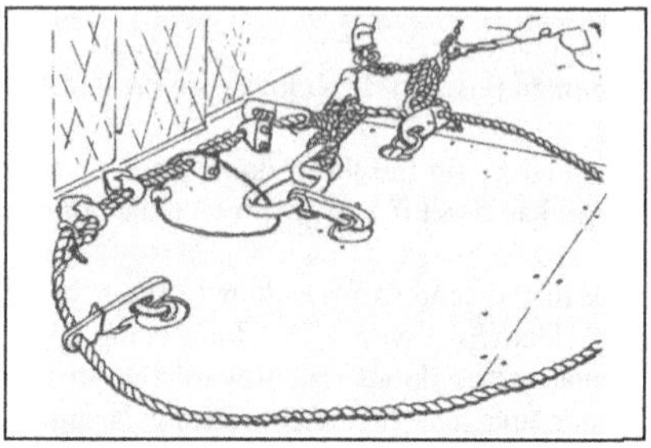

Figure 3-3. Donut ring and rappel rope connection in a helicopter.

(1) Thread six parachute static line snaphooks onto the 120-inch steel cable so that four snaphooks are facing out with the gates down, and the center two snaphooks are facing in with the gates down. Drill the end portion of each snaphook to make a 5/8-inch diameter hole, and thread the cable through the holes in the snaphooks.

(2) Overlap the ends of the 120-inch cable 20 inches to form a circle. Secure the ends with four 1/2-inch U-bolts placed at 2- to 3-inch intervals.

(3) Attach two U-bolts to each dead end of the cable so that the bolts engage the dead end.

(a) Before torquing the U-bolts, position a 12-inch length of chain or 1/8-inch diameter cable on the center of the overlapped 120-inch steel cable so that it remains in position between the two center U-bolts.

(b) Tighten each nut of the U-bolts with a torque wrench (if possible) to 40-foot pounds (480-inch pounds).

(c) After the U-bolt clamps have been attached and tightened, fasten a steel plate (drill to fit) over the open end of the U-bolt studs and spot weld in place to prevent loosening.

b. **Floating Safety Ring**. The floating safety ring is referred to as the secondary (or No. 2) anchor point for the rappelling ropes (Figure 3-4). The snaplink at the end of the rappelling rope is hooked to this connection. Either of the following two types of floating safety rings may be used:

(1) *First method.* Thread an elliptical rappelling ring through the free end of the keeper chain (cable). It is constructed of cold-rolled steel that is 1/4-inch in diameter, with inside dimensions of 2 1/4 inches (minor axis) and 4 inches (major axis). Thread a seventh parachute static line snaphook onto the ring before welding. Weld this ring together so that it withstands a force up to 3,000 pounds.

(2) *Second method.* Attach two snaplinks to the aircraft tie-down ring in the center of the donut ring. Insert the first snaplink through the free end of the keeper chain (cable) and the tie-down ring with the gate down. Insert the second snaplink through the free end of the keeper chain (cable) and the tie-down ring with the gate up. Tape the snaplink

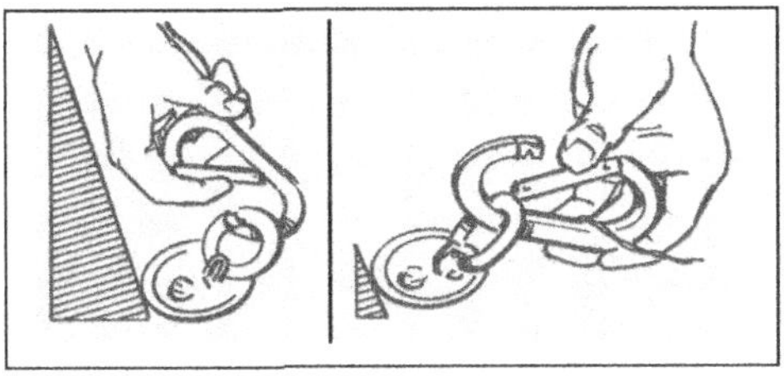

gates closed with masking tape. Then, tape the snaplinks together to ensure that the snaplink gates are on opposite sides of each other.

Figure 3-4. Floating safety ring formed with two snaplinks.

c. **Rappel Rope Anchor Points.** The rappelling rope is connected to the floating safety ring and the donut ring as follows:

(1) *Number 1 anchor point (donut ring).* Using a bight about 5 feet from the end of the standing part of the rope, make one turn through the snaplink forming a round turn. The bowline is the preferred method of attaching rappel ropes to snaplinks or anchor systems. The round turn with two half hitches is a reliable means of attaching ropes to anchor systems. Secure the round turn to the snaplink with two half hitches. Make the connection to the donut ring by attaching the snaplink gate upward. Ensure the gate faces upward with the opening away from the knot.

(2) *Number 2 anchor point (floating ring).* Attach the No. 2 snaplink the same as the first with some exceptions. Using a bight about 2 feet from the end of the standing part of the rope, connect the snaplink to the rope the same as the first connection. Tape the end of the standing part of the rope and the knots with masking tape (or green engineer tape) to secure them in place. Make the connection to the floating safety ring the same as the connection to the donut ring (Figure 3-5). Four ropes can be connected to the floating safety ring using two snaplinks (Figure 3-6).

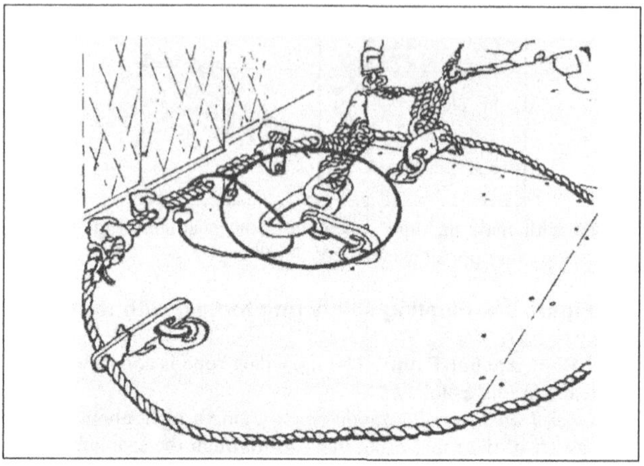

Figure 3-5. Rappel rope connection using two snaplinks for the floating safety ring.

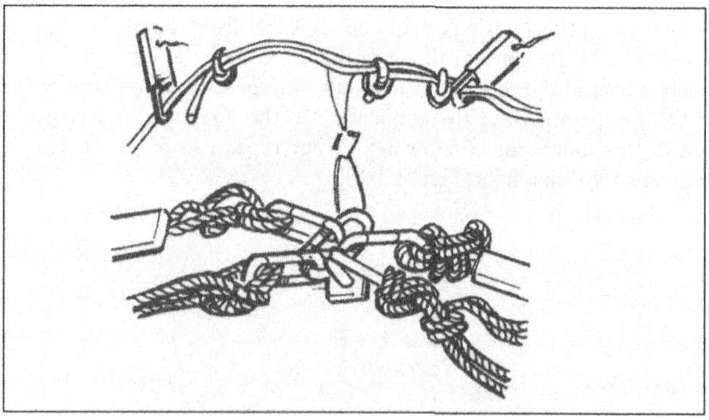

Figure 3-6. Four rappel ropes connected to the floating safety ring (two snaplinks).

3-18. SEATING ARRANGEMENTS AND LOADING TECHNIQUES

Rappelling operations are executed quickly and safely to ensure the success of the operation in a combat or training environment. All rappellers must know the seating arrangements and the procedures for loading (Figure 3-7). They must be rehearsed under the supervision of the rappel master before conducting the operation.

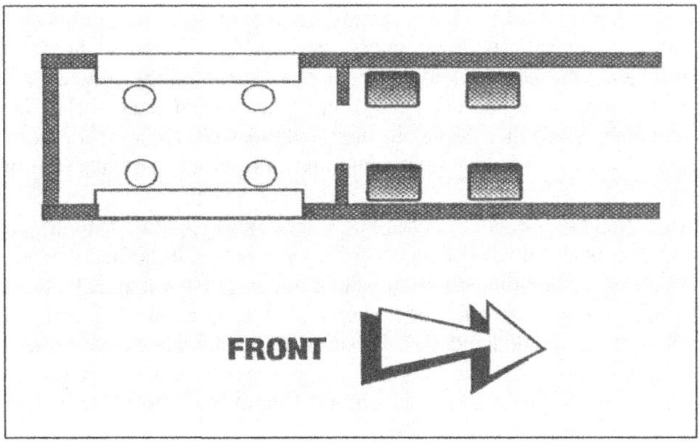

Figure 3-7. Seating arrangement.

a. The first four rappellers are hooked up to the donut ring and use safety belts while in flight. The remaining rappellers use seat belts or safety straps. For safety emphasis during flight, each rappeller's brake hand is in position with no slack in the rope between the brake hand and the donut ring. The rappeller ensures that his rappelling rope is not tangled on any part of the interior of the helicopter or his equipment. He ensures that the coiling log or D-bag is located in such a position that he drops the rope with ease, using his guide hand. The coiling log or D-bag is secured with half of it under the rappeller's inside leg and the other half on top of his outside leg.

b. The rappellers should board the aircraft and sit along the leading edge of each door. They hook up with the rappel seat square knot on the inboard side of the aircraft and the brake hand on the outboard side. The rappel master sits in the center to aid visual inspection of personnel and equipment.

c. For a flight time of less than five minutes and for training and operations where speed is a factor, the rappel master directs the rappellers to be seated in the door facing outward with their feet on the skid. In a tactical situation and before the helicopter comes to a hover, the rappellers assume an upright position facing inside the helicopter.

d. The aircraft is cleared as fast as possible. This reduces the time that the aircraft is in the high-hover attitude to a minimum.

3-19. RAPPELLING PROCEDURES

The rappel master ensures all personnel follow the rappelling procedures for the safe and efficient execution of the operation. These procedures are rehearsed under the supervision of the rappel master.

a. Rappellers (eight maximum) are identified, and rappel masters tell them which door of the aircraft to exit. Rappellers approach the aircraft once it is firmly on the ground, and the crew chief signals when it is safe to approach. The approach to the aircraft should be about 45 degrees to the nose of the aircraft allowing the pilot to see the rappeller.

b. If the rappel master is not already on the aircraft, he boards first and secures himself to the aircraft with a safety harness. He dons a headset to maintain communications with the pilot and positions himself behind the crew chief's chair.

c. As the rappellers approach the aircraft, they hand their anchor attaching points (snaplinks connected to their ropes) to the rappel master. The rappel master connects one set of snaplinks to the floating safety ring and one set of snaplinks (the primary) to the donut ring.

d. All rappellers use safety belts. For safety emphasis during flight, each rappeller's brake hand is in position with no slack in the rope between the brake hand and the donut ring. Rappellers ensure that their rappelling ropes are not tangled on any part of the interior of the helicopter or their equipment. They ensure that they can drop their D-bag or log coil using the guide hand. The D-bag or log coil is placed so it does not fall out of the aircraft. Once over the target area, the rappel master issues the rappel commands and ensures the rappel ropes are on the ground (about 20 feet of rope should be on the ground).

e. On the command GO, the rappeller flexes at the knees and vigorously pushes away from the skid gear, allowing the rope to pass through his brake hand and guide hand. The descent is accomplished smoothly at a rate of about 8 feet per second, avoiding jerky stops. The rappeller initiates his braking action slowly when he is about halfway to the ground. The safety officer or NCO ensures that at least a five-second delay is maintained between each rappeller. Another technique used to deploy rappellers is to issue the command GO to the rappellers diagonally opposite each other. In the early phase of performance, the rappeller's feet are together (particularly if over woods or jungle) and his guide hand is used (not the brake hand) to remove any entanglements.

f. Upon reaching the ground, the rappeller clears the rappel rope through the snaplink (or rappel ring) until the rope is free. If other rappellers are to follow on the same rope, the rappeller on the ground separates or untwists the ropes and becomes the belayer for each subsequent rappeller. If ropes are to be released from the aircraft following the last rappeller, untwisting and separating the ropes is not necessary. A two-second interval is maintained between exit groups (two men exiting at the same time).

g. The rappel master releases the rope from the donut ring after he confirms (by visual inspection) that the rappeller is off rappel. He then drops the rope away from the helicopter.

3-20. RAPPELLING COMMANDS

Helicopter rappelling is conducted in a noise-filled environment. Each rappeller is trained to know the rappel commands and to understand the actions required of him during the execution of each command. Due to the noise created by the helicopter, all rappel commands should be accompanied by arm-and-hand signals. All commands are issued by the rappel master and confirmed orally by the rappellers.

a. The helicopter rappelling commands are as follows:

(1) GET READY. The rappeller checks his combat equipment, looks toward the donut ring, and pulls the rope to check the anchor point connection. He then checks his

snaplink to ensure the rope is properly seated in the snaplink on his rappel seat.

(2) THROW ROPE. The rappeller drops his deployment bag out and away from the helicopter with his guide hand. He ensures that the rope does *not* fall between the side of the helicopter and the skid. The rappel master ensures the rope is touching the ground and is free of tangles and knots. The rappeller then replies, "Rope okay."

(3) SIT IN THE DOOR. The rappeller swings his legs to the outside of the helicopter and takes up a sitting position. (This command applies only to the UH-1H helicopter.)

(4) POSITION. The rappeller pivots 180 degrees on the skid. His feet are shoulder-width apart, the balls of his feet are on the skid, his knees are locked, his body is bent forward in an L-shape position, and his brake hand is on his buttock. The rappel master makes a visual inspection of the snaplink and rappel ring.

(5) GO. The rappel master points at the rappeller to exit. One rappeller from each side rappels at the same time. The rappel master gives the command GO to the rappellers that are diagonal to each other—for example, the front left rappeller and the rear right rappeller. The rappeller flexes his knees and jumps backward, letting the rope run through the brake hand and guide hand.

b. The arm-and-hand signals that are used with rappelling commands are as follows:

(1) GET READY (Figure 3-8). The rappel master extends both arms to the front with fists clenched and thumbs pointing upward.

Figure 3-8. Arm-and-hand signal for GET READY.

(2) SIT IN THE DOOR (Figure 3-9). The rappel master extends both arms to the front with elbows locked, fingers extended, thumbs running along the index fingers, and palms facing downward. He bends slightly at the waist so that the arms are below the waist. Then, he moves both arms in a crisscross waving motion, alternating left over right and right over left.

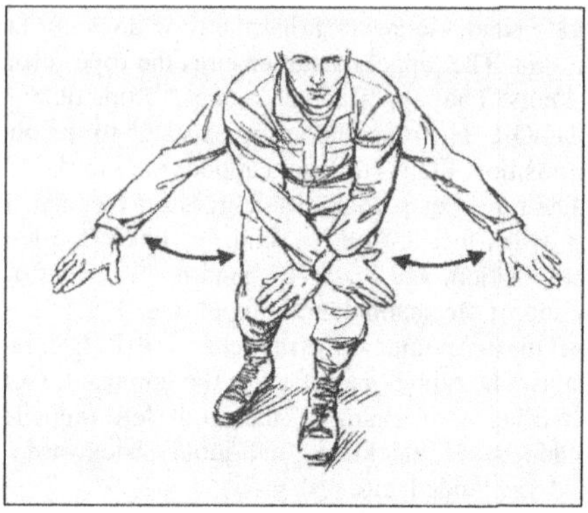

Figure 3-9. Arm-and-hand signal for SIT IN THE DOOR.

(3) THROW ROPES (Figure 3-10). The rappel master extends both arms to the front with elbows locked, fist clenched and index fingers extended. He points at respective rappellers, brings the forearms to an upright position and then down to the elbow-locked position.

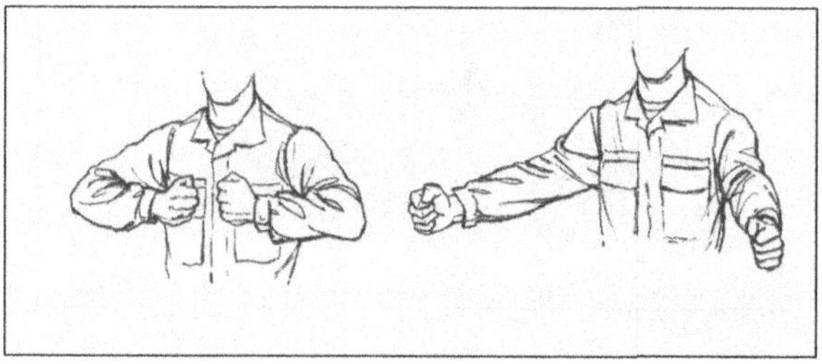

Figure 3-10. Arm-and-hand signal for THROW ROPES.

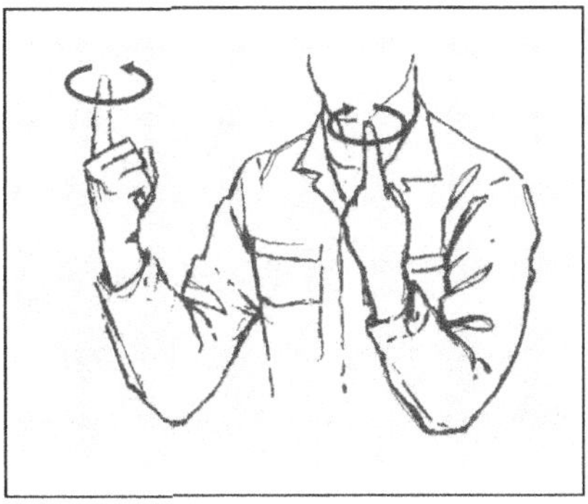

Figure 3-11. Primary arm-and-hand signal for POSITION.

(4) POSITION. The two arm-and-hand signals for POSITION are:

(a) Primary signal (Figure 3-11). The rappel master extends both arms to the front, elbows bent, forearms pointed upward, fists clinched, and index fingers pointing upward. He makes a circular motion with both forearms rotating in opposite directions.

(b) Alternate signal (Figure 3-12). The rappel master extends both arms to the front with elbows locked, fists clenched, and index fingers extended. He bends at the waist so that his arms are below his waist and makes a circular motion with arms rotating in opposite directions.

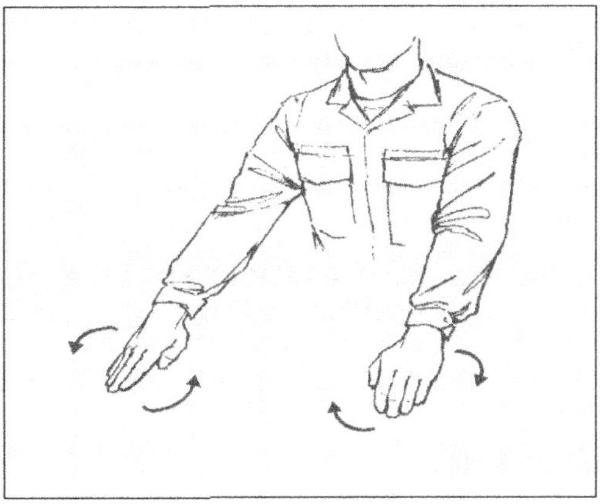

Figure 3-12. Alternate arm-and-hand signal for POSITION.

(5) GO (Figure 3-13). The rappel master extends an arm with elbow locked, fingers extended, and thumb along the index finger, and he points directly at the rappeller.

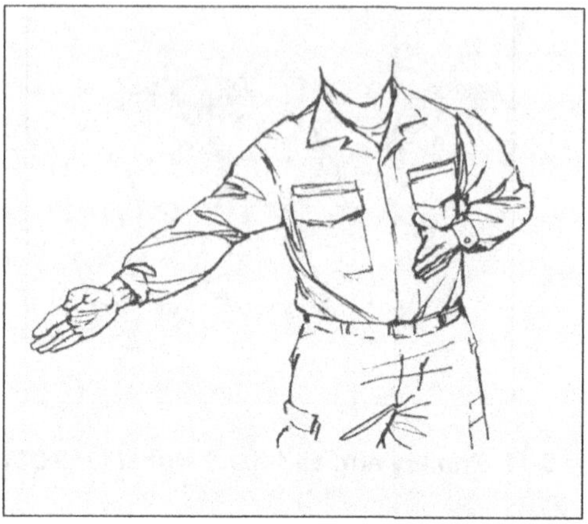

Figure 3-13. Arm-and-hand signal for GO.

3-21. INSPECTION AND SAFETY CONSIDERATIONS

The rappel master and pilot (or his representative) conduct a joint inspection of the aircraft to ensure the safety of all personnel and serviceability of equipment.

a. Cargo doors are locked in the open position or cleared for closing, depending on the mission.

b. All loose objects in the cargo compartment are removed or secured forward.

c. Sharp edges or protrusions on the cargo floor and door ledges are taped.

d. The donut ring and floating safety ring are serviceable and properly attached to tie-down fittings.

e. A headset and intercom jack for the rappel master are available and operational.

f. Serviceable safety harnesses are available for the rappel master and crew chief.

g. The recovery rope is installed and properly stowed.

h. Unused floor rings are taped down.

Section VI. RAPPELLING OPERATIONS FOR UH-60 BLACKHAWK HELICOPTER

The techniques used by the rappeller when rappelling from different aircraft are similar. However, positioning, seating, and the tie-down anchor point are different. Each rappeller is well trained on each aircraft before conducting a rappel.

3-22. CHARACTERISTICS

The UH-60 is a twin-engine, medium-speed, single-main rotor helicopter that can transport 10 rappellers, 1 rappel master, and 1 four-man crew when the center row of troop seats is removed. It is equipped with four 4,000-pound (load limit) cabin ceiling tie-down fittings or rings that are located in the ceiling of the troop/cargo compartment. These fittings are used as the primary rappelling rope anchoring points. The UH-60 is also equipped with eight 3,500-pound (load limit) cargo restraint net rings. Four of the eight rings are located in the ceiling of the troop/cargo compartment. These rings are used as the secondary rappelling rope anchoring points.

3-23. RIGGING OF UH-60 FOR RAPPELLING

The rappel master rigs the UH-60 helicopter by performing the following:

a. Locks both cargo doors in the open position.

Note: For arctic or other cold-weather operations, or during flights of long duration, the cargo doors are closed and locked until the time specified for opening them.

b. Removes the center row of troop seats.

c. Tapes any sharp edges or protrusions on the cargo floor and door ledges that may come in contact with the rappeller or the rappelling rope.

d. Stows loose equipment forward in the cargo compartment.

e. Extends the rappel master's intercom cord to the rear over the aft utility drain line and tapes the cord to the overhead troop seat support tube.

f. Installs the floor restraint provisions for rappellers No. 1 through 6.

g. Rigs and connects rappelling ropes to the aircraft's primary and secondary anchoring points.

(1) *Primary anchor points.* Ties a bowline with a half hitch about 4 feet from the standing end of the rope. Attaches the two primary snaplinks to the respective cabin tie-down fitting ring with gates facing in the opposite directions (Figures 3-14 and 3-16)

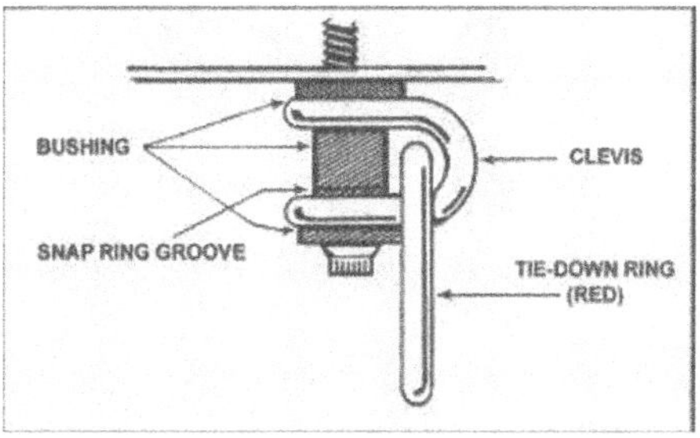

Figure 3-14. Primary snaplink attaching point.

(2) *Secondary anchor points.* Ties a bowline with a half hitch about 1 1/2 feet from the standing end of the rope. Attaches the secondary snaplink to the adjacent overhead cargo restraint net ring (Figures 3-15 and 3-16).

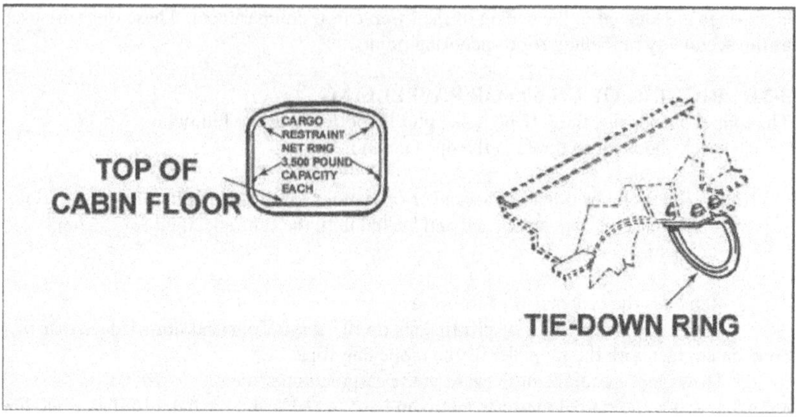

Figure 3-15. Secondary snaplink attaching point.

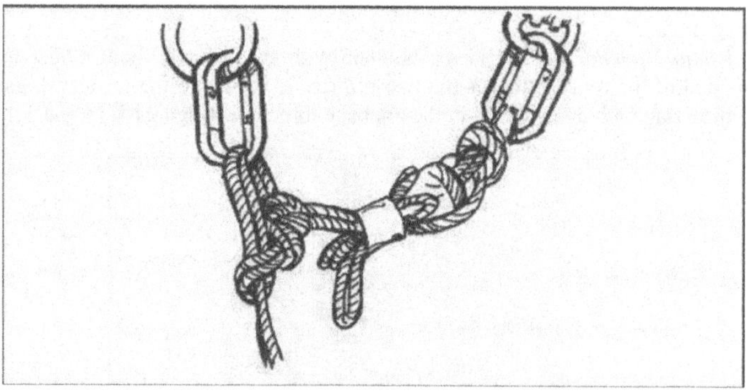

Figure 3-16. Primary and secondary snaplink attaching points.

h. Removes and secures the cargo hook access door and deploys the cargo hook in the DOWN position.

i. Installs the recovery rope for endangered rappellers.

j. Tapes the unused floor rings.

Note: There must be a minimum of 22 inches and a maximum of 30 inches between the anchor knots.

3-24. SEATING ARRANGEMENTS AND LOADING TECHNIQUES

Rappelling operations are executed quickly and safely to ensure the success of the operation in a combat or training environment. All rappellers must know the seating arrangements and the procedures for loading. They must be rehearsed under the supervision of the rappel master before conducting the operation.

a. A maximum of 10 rappellers, with and without combat equipment, and 1 static rappel master is seated and restrained aboard the UH-60 helicopter when the center row of troop seats are removed (Figure 3-17).

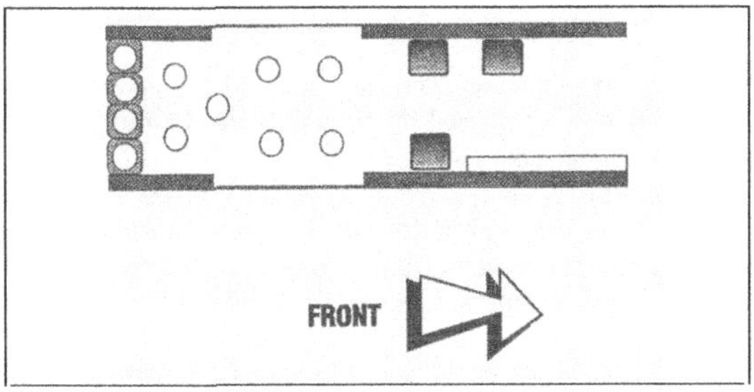

Figure 3-17. Rappelling personnel seating arrangement.

(1) Rappellers No. 1 through 6 and the rappel master are seated on the cargo floor. The rappellers are restrained using 3 1/2 feet of 1-inch tubular-nylon webbing (or equivalent) that is run through the cargo tie-down rings attached to the floor of the aircraft. The webbing is attached to the rappeller by means of a snaplink that is hooked to a Swiss seat. The rappel master is secured by his safety harness.

(2) Rappellers No. 7 through 10 are seated on the troop seats across the aft end of the cargo compartment. They are secured by the seat belts.

b. Rappellers are organized into 10-man sticks. They approach the aircraft from the left or right side in reverse order. They are divided into two groups and approach and enter the aircraft at the same time, depending on the situation. Rappellers No. 9, 7, 5, 3, and 1 enter the aircraft through the left door; they are seated and restrained. Rappellers No. 10, 8, 6, 4, and 2 enter the aircraft through the right door; they are seated and restrained. The static rappel master is restrained in the center of the cargo compartment using a C-3A troop-type safety belt or a safety harness during take-off and landing.

DANGER

BECAUSE THE LOWEST ARC OF THE ROTOR BLADE OCCURS AT THE DIRECT FRONT OF THE AIRCRAFT, APPROACHING THE AIRCRAFT AT THIS POINT COULD RESULT IN PERSONNEL INJURY OR DEATH.

3-25. RAPPELLING PROCEDURES

The rappel master ensures all personnel follow the rappelling procedures for the safe and efficient execution of the operation. These procedures are rehearsed under the supervision of the rappel master.

a. The rappel master is equipped with a headset and maintains communications with the pilot at all times. He wears a safety harness and stations himself in the center of the cargo floor to maintain control of the rappellers, their ropes, and all anchor attaching points.

Note: If all rappellers are rappelling into the same area, individual rappel ropes are not needed. The ropes used by rappellers No. 1 through 4 are used by the succeeding rappellers.

b. After all rappellers are in the aircraft, rappellers No. 1 through 4 hand the rappel master their primary and secondary anchor point snaplinks. The rappel master hooks the primary anchor point snaplink into the respective cabin tie-down fittings and the secondary anchor point snaplink into the respective cargo restraint net rings. After the primary and secondary anchor point snaplinks are secured, the first four rappellers hook onto their respective rappel rope. The rappel master checks the first four rappellers for the correct hookup.

c. After the first four rappellers exit and clear their ropes, the rappel master hands the next four rappellers their rappel ropes to hook up. Left-handed rappellers take extra precaution to ensure the correct hookup. Problems can be avoided by placing left-handed rappellers in the positions where they initially have the rope to their nonbrake side.

WARNING

Extreme care must be taken in hooking up and positioning personnel that require repositioning of their rope to their proper brake-hand side. This could result in bodily injury or damage to Army property.

d. After checking the rappellers for the correct hookup, the rappel master moves the next four rappellers into position on the floor of the aircraft. After the four rappellers exit and clear their ropes, the rappel master hands the last two rappellers their rappel ropes to hook up.

e. After checking the rappellers for correct hookup, the rappel master moves them into position on the floor of the aircraft. After the last rappeller is off the rope, the rappel master releases the ropes.

f. If the rappel master rappels, the crew chief is instructed to release or retrieve the ropes.

3-26. RAPPELLING COMMANDS

The UH-60 rappelling is conducted similar to the UH-1H rappelling with the exception that no skids are on the UH-60 on which to stand. Therefore, the edge of the floor, along the door of the helicopter, is used as a pivot point to assume the L-shaped rappelling position.

a. GET READY. This alerts the rappeller. The rappeller and rappel master perform final checks of the hookup, rappel seat, snaplink, and equipment.

b. THROW ROPES. The rappeller looks down below the aircraft to ensure that no one is under the aircraft. Keeping his brake hand in the small of his back, he tosses the deployment bag with the ropes *out and away* from the helicopter with his guide hand. The rappeller observes that the ropes are touching the ground and are not knotted or entangled.

c. POSITION. From a kneeling position with the brake hand in the small of the back, the rappeller rotates 90 degrees so that he is facing the inside of the aircraft and the rappel master. The rappeller then places his heels on the edge of the floor of the helicopter doorway and leans out into an L-shaped position. His feet are shoulder-width apart, the balls of his feet are on the edge of the helicopter doorway, his knees are locked, and his body is bent at the waist toward the helicopter.

d. GO. The rappel master initiates the rappel with this command. The rappeller flexes his knees and jumps backwards. At the same time, the rappeller throws his brake hand out at a 45-degree angle, letting the running ends of the ropes slide through both the brake hand and guide hand.

3-27. INSPECTION AND SAFETY CONSIDERATIONS

The rappel master and pilot (or his representative) conduct a joint inspection of the aircraft to ensure the safety of all personnel and serviceability of equipment.

a. Cargo doors are locked in the open position or cleared for closing, depending on the mission.

b. All loose objects in the cargo compartment are removed or secured forward.

c. Sharp edges or protrusions on the cargo floor and door ledges that may come in contact with the rappeller or his rappelling rope are taped.

Note: Do not tape the door latches or handles; this can interfere with door operations.

d. Primary and secondary rappelling anchor points are serviceable and securely attached to the aircraft structure.

e. A headset/helmet and intercom jack for the rappel master are available and operational, and the intercom extension cord is secured overhead.

f. Serviceable safety harnesses are available for the rappel master and crew chief.

g. The cabin ceiling tie-downs must have safety wire installed to ensure they do not come undone or unraveled, and the bolt head must be stamped H.

Section VII. RAPPELLING OPERATIONS FOR MH-53 HELICOPTER

The MH-53 helicopter is a highly versatile aircraft. It is used for a variety of missions, usually in conjunction with special operations.

3-28. CHARACTERISTICS

The MH-53 is a two-engine, single-rotor, heavy-lift helicopter. It has a crew of six and can be refueled while in flight. It has a precision navigational and communications package, which makes it excellent for conducting deep infiltration and exfiltration missions. The MH-53 helicopter can fly at night in all weather and terrain conditions. It can follow down to 100 feet and operate from unprepared sites. It is fitted with three gun stations for a mix of 7.62-mm miniguns and caliber .50 machine guns.

3-29. RIGGING OF MH-53 HELICOPTER FOR RAPPELLING

When rappelling from the crew entrance door, rappellers connect the ropes with locking snaplinks to the two 10,000-pound tie-down rings under the left scanner's window. When rappelling from the ramp, rappellers connect the ropes with locking snaplinks to the two 10,000-pound tie-down fittings on the left and right sides of the aircraft at station 522. Any sharp edges that could damage the ropes should be padded or taped. A length of fire hose may also be used over the portion of the rope that comes in contacts with the door or ramp edge.

3-30. SEATING ARRANGEMENTS AND LOADING TECHNIQUES

The MH-53 is a large, versatile helicopter. The seating arrangements and loading techniques are numerous. The using unit develops a seating and loading SOP that ensures the safety of all personnel and permits smooth, efficient execution of the mission.

3-31. RAPPELLING PROCEDURES

The rappel master or his assistant are at the two exit points (ramp and door). The rappel master maintains communications with the helicopter commander and relays all commands and time warnings. The commander issues time warnings at 20-minute, 10-minute, 5-minute, and 1-minute intervals. During limited visibility, the rappel master may use NVDs to observe the safety/belayer. Chemical lights are attached to the top or bottom of the rappel rope. Arm-and-hand signals should be IAW The Air Assault School Handbook and FC 57-2.

Note: Night vision goggles cause limited depth perception and a tunnel vision effect.

a. The rappel master is secured by a harness and ensures the proper hook up of each rappeller. Once hooked up, the rappellers release their safety belts. At the command of the rappel master, the rappellers position themselves to ease immediate deployment.

WARNING

Before conducting deployments off the ramp, the deploying team members must be briefed on the importance of maintaining separation between members. About 24 to 27 inches are desired. This separation helps maintain aircraft center of gravity limits.

b. The rappel master deploys ropes only after the helicopter is in a stable hover over the target area and he has given the command ROPES. He ensures the ropes are on the ground before giving the command GO. After the last rappeller is off the rope, the rappel master retrieves or cuts the ropes away.

3-32. RAPPELLING COMMANDS

The helicopter rappelling commands for the MH-53 are as follows:

a. GET READY. This command is given as the helicopter approaches the rappel site. It alerts the rappellers of the approach to the site and signals them for a final equipment check.

b. POSITION. This command clears the first rappellers into position for deployment.

c. THROW ROPE. This command is given once the helicopter reaches a stable hover.

d. GO. The rappellers exit.

3-33. INSPECTION AND SAFETY CONSIDERATIONS

The rappel master briefs all personnel participating in the operation. He conducts hands-on inspection to ensure the safety of all personnel and serviceability of equipment.

a. Tie-down fittings are serviceable.

b. All sharp or protruding edges that may come in contact with the rappelling ropes are padded or taped.

c. All ropes are retrieved or cut away before forward movement of the helicopter.

d. Only three rappellers are deployed at a time—two from the ramp and one from the personnel door.

CHAPTER 4

SPECIAL PATROL INFILTRATION/EXFILTRATION SYSTEM

The special patrol infiltration/exfiltration system (SPIES) provides an excellent form of exfiltration over short distances. It is not recommended for infiltration because team members are exposed the entire time. Due to the nature of SPIES operations, a thorough briefing is required for all participants before the operation. Personnel being extracted must train extensively in the SPIES extraction before infiltration. For the other personnel involved, a complete preoperations briefing is held before the operation. This is crucial when additional assets are involved other than the extraction helicopter such as gunships, aerial observers, or artillery support.

Section I. SPIES MASTER DUTIES AND QUALIFICATIONS

The SPIES master holds the rank of sergeant or above (may be waived by the commanding officer). The commanding officer ensures that this qualification is entered on the soldier's training record.

4-1. QUALIFICATIONS

To qualify to be a SPIES master, the sergeant must participate in at least three SPIES operations (observe twice and execute SPIES master duties once under observation). For example, he hooks up the helicopter, assists in the preparation of an operation, and conducts a successful operation under the supervision of a qualified SPIES master. A SPIES master is conscientious and thorough, and he knows all aspects of a SPIES operation. He can give an effective pilot's brief, use the aircraft communications equipment, and understand aviation terminology.

4-2. DUTIES

The SPIES master is responsible for the safe conduct of the SPIES operation.

a. **Preflight Duties.**

(1) Inventories and inspects all SPIES equipment.

(2) Briefs the pilot and other concerned personnel about details of the operation, especially the extraction and dismounting procedures.

(3) Ensures that an ICS helmet and gunner's belt are available for the SPIES master's use. Uses a sling rope if no belt is available. Connects and checks the operation of the ICS to be used. (ICS communications is established between the SPIES master and pilots on all SPIES operations.)

(4) Attaches the SPIES rope to the helicopter IAW the guidance in this chapter.

(5) Ensures that nothing is adrift in the aircraft that may fall on a team member.

(6) Checks the location of the emergency axe. Places the axe where readily available, yet secure enough so as not to endanger the men on the SPIES rope. (Inspects the axe to ensure that it is sharp.)

b. **Extraction Duties.**

(1) On arrival at the team's estimated position, assists the pilot in determining the exact location of the team members.

(2) As the aircraft approaches the team's location, aids the pilot (using the clock system) in placing the aircraft directly above the team.

(3) Requests permission from the pilot to drop the SPIES rope when the aircraft is hovering above the team.

(4) Drops the rope, taking care to avoid striking team members on the ground.

(5) Notifies the pilot when the rope is down, and reports all altitude corrections to ensure that team members reach all SPIES attachment points.

(6) Watches for the "thumbs-up" signal from all team members.

(7) On receipt of the "thumbs-up" signal, advises the pilot that the team is ready for extraction and requests a vertical lift-off.

(8) Advises the pilot of the team's position, the location of any potential obstacles, and the avoidance of horizontal movement.

(9) If a team member becomes entangled with an obstacle during the extraction, notifies the pilot and requests that the vertical lift be stopped. If the situation is critical, prepares to cut the SPIES rope (the anchor point or cargo straps) after the team members are secured to the obstacle or on the ground.

(10) When positive that all obstructions are clear, advises the pilot to obtain a safe altitude (about 350 feet AGL for training purposes or as the situation dictates in combat) or to transition into forward flight.

(11) At frequent intervals during the flight, advises the pilot on the safety status of all team members. Maintains a constant visual watch on the team and checks security of the SPIES attachments often.

c. **Dismounting Duties.**

(1) On arrival at the dismounting area, informs the pilot as to the approximate distance of the lower rope end from the ground.

(2) Once the pilot starts the vertical descent, continually informs him as to the approximate distance of the lower rope end from the ground.

(3) Informs the pilot of any horizontal drift that occurs and any obstructions near the SPIES rope. Also informs the pilot of any oscillation that may occur.

(4) Informs the pilot when the rope is about 25 feet above the ground and again when it is 10 feet above the ground. Ensures that the rate of descent is slow enough to enable the team members to land and get out from under team members safely.

(5) Reports when the first man initially touches down, when the last team member starts to safely move away from under the helicopter, and when all team members are disconnected.

(6) On order of the pilot, either retrieves the SPIES rope into the helicopter or disconnects the SPIES rope and drops it to the ground. When using the UH-1H/UH-60 helicopter, the only way to retrieve the SPIES rope while in the air is by having an arranged recovery rope attached with a 16-foot sling rope. In some cases, the SPIES master joins two 12-foot-long sling ropes to haul the SPIES rope aboard and attaches the rope about 5 or 6 feet below the cargo hook or cargo strap hookup point. The type of knot used to connect the sling (or recovery) rope to the SPIES rope is self-tightening in nature (for example, the prusik knot). The SPIES master fastens the standing end of the sling rope to the deck tie-down or uses a snaplink. Although it is important to keep the line out of the way, the primary consideration is its length. The rope must be long enough for any oscillation in the SPIES during flight.

Section II. PREOPERATIONS BRIEFINGS AND PROCEDURES

Before conducting a SPIES training mission, the participants must have a basic understanding of the requirements. The SPIES master conducts briefings to ensure the soldier extracted and the pilot know the procedures. As SOPs are developed and units train together, the SPIES master simply refers to the SOP. He always gives a safety briefing and conducts an equipment inspection.

Note: SPIES has not been approved for Army-wide use. Approval must be obtained from division level or higher commander before the conduct of any training.

4-3. SAFETY

As in all training, a safety briefing must precede operations using the SPIES. The briefing should consist of, but not be restricted to, a review of the following:

- Characteristics of equipment associated with the SPIES.
- Equipment inspection.
- Proper donning of the harness.
- Methods of extraction and insertion used.
- Emergency signals.

4-4. COMMUNICATIONS

Due to the noise level associated in all helicopter operations, radios are used for communication. They are also used before the arrival of the helicopter. Precise arm-and-hand signals are established in case of radio failure or poor communications due to static or noise overriding the audio output of the radio. During the first part of the operations, the SPIES master either observes (daytime) or knows that a definite procedure is taking place (night or jungle) while the teams are hooking up to the SPIES rope for extraction. Headsets and voice suppressors replace the handset for better radio procedures. This allows the RATELO on the ground to use both hands while the helicopter hovers for a faster and safer hookup. If radio communications are hampered, special procedures are used along with hand or light signals.

4-5. EXTRACTION PROCEDURES

After the team has been located, the SPIES master assists the pilot in directing the helicopter to the proper distance over the team.

a. The team leader positions himself so that he can move and approach the rope as it is dropped by the SPIES master. Once the rope is clear of any obstacles, the team leader signals the team to their assigned positions along the 10 hookup points. Using the primary or harness snaplink, each team member hooks to the D-ring on his side of the line. This is the primary hookup. Once the team is hooked up, the team leader hooks into the alternate or second hookup point using the safety line and snaplink. He then faces forward along the line so that he is heading in the direction he is traveling when the aircraft starts its ascent.

b. The SPIES rope is held up and routed over the shoulder closest to the rope. With the other hand, the team members give thumbs-up signals to allow both the team leader and the SPIES master to see that they are ready to go. The team leader physically inspects (if time and situation permits) or hooks himself in on the lowest point along with the RATELO to ensure the running end is clear of all obstacles. He gives the thumbs-up signal to the SPIES master. This thumbs-up signal (at night, an arranged light signal) continues until the team is lifted off the extraction surface. The team leader also signals (arms held straight to the sides) when the team is at safe altitude (approximate 10 feet above the tallest obstacle at the extraction site). The helicopter starts a transition in a horizontal direction on its return flight.

4-6. EMERGENCY PROCEDURES

During the flight, from extraction until the team is safely and quickly detached from the SPIES rope, each team member should be aware of any problems that arise from above or below. The man above checks the man below. At the first sign of danger or if an emergency arises, the team leader or a team member places his free hand on his head. Upon observing anyone on the SPIES rope with his hand on his head, the SPIES master tells the pilot to make an emergency landing in the nearest and safest area.

4-7. DISMOUNTING PROCEDURES

During the familiarization training phase, the SPIES master ensures all members know they should immediately head in the direction of the nose (12 o'clock) of the aircraft. This allows the pilot to see that the team is out from under the aircraft. When an emergency situation arises, the pilot can make a better appraisal of the situation if he can see all the members of the team. When the helicopter is making a scheduled landing, the team ensures that the SPIES rope does not interfere with the aircraft and that the aircraft does not land on the rope.

4-8. INSPECTION OF EQUIPMENT

The SPIES is inspected by a certified SPIES master at six-month intervals and whenever the SPIES master questions serviceability. Outdated, spliced, abraded, or cut rope is removed from service. However, loose or broken stitching in excess of three stitches is not repaired.

CAUTION

Acid contamination, cuts, or fraying of harness or sling webbing constitute nonrepairable damage.

a. The certified SPIES master inspects the harness and suspension sling webbing for signs of contamination from oil, grease, acid, or rust. He also checks for cuts, twists, fading, excessive wear, fusing (indicated by unusual hardening or softening of webbing fibers), fraying, burns, abrasions, and loose or broken stitching. The damaged harness or suspension sling is removed from service and returned to supply for appropriate disposition.

b. The SPIES master inspects all hardware for signs of corrosion, pitting, ease of operation, security of attachment, bends, dents, nicks, burrs, and sharp edges.

Replacement of hardware (except chest strap adapter) that requires unstitching of webbing renders the harness unserviceable.

c. The SPIES master replaces the V-ring by cutting the strap above the stitching. He folds and stitches a new end section of the leg strap. If the leg strap is damaged, the SPIES master returns the harness or suspension sling to supply for appropriate disposition.

d. The SPIES master checks ropes, harnesses, and suspension slings for expiration of service or total life, which is 7 years of service (opening manufacturer's package) or 15 years from date of manufacture, whichever occurs first. The SPIES master also ensures the rope is free of splices. He ensures the eye loop at the end of the SPIES rope is not broken, frayed, or loose.

e. The SPIES master inspects the rope surface for cuts, excessive abrasions, and snags. Cuts on the rope are excessive when there are four or more cut strands in any 5-inch length. The two-to-one braided rope has 12 pairs of strands (24 individual strands) around the circumference. Abrasion is extensive when torn yarns are equivalent to that of four strands of any 5-inch length. Rope that has been subjected to heavy loads may display glazed areas where it has worked against hard surfaces. This condition may be caused by paint or fused fibers. After long use, the rope may become fuzzy on the surface (although this should be minimized with the surface coating). However, the effect on the rope's strength is negligible.

f. The SPIES master inspects the rope for signs of contamination by acid, alkaline compounds, salt water, fire extinguishing solutions, and petroleum-based solvents. Although the ropes gradually change color, such changes do not indicate decrease in strength unless the change is due to contact with strong chemicals. Changes in color caused by chemicals are usually spotty; changes caused by use are uniform throughout the length of the rope.

4-9. RIGGING OF UH-1H HELICOPTER FOR SPIES OPERATION

Personnel who are not familiar with SPIES are encouraged to observe or take part in the rigging of the helicopter. This builds confidence in the equipment and assists in a more comprehensive training of new SPIES masters.

a. **Equipment.**

- One 120-foot SPIES rope with deployment bag.
- Two 11-foot, 3- or 4-loop cargo slings or two 9-foot, 3- or 4-loop cargo slings.
- Two Type IV connector links (four if aircraft does not have cargo hooks).
- Heavy duty tape (100-mph tape).
- One 12-foot sling rope.
- Five oval snaplinks (nine if the aircraft does not have cargo hooks).

NOTE: If the aircraft does not have cargo hooks, use four 11-foot or four 9-foot, 3- or 4-loop cargo slings.

b. **Installation.**

(1) The primary attachment point for the SPIES rope is the cargo hook. The end of the SPIES rope having a polyurethane encapsulated eye is attached to the cargo hook.

The two 9- or 11-foot long cargo suspension slings are joined to form one continuous sling, using a Type IV link. The sling is stretched out on the helicopter deck. One end is taken under the helicopter and through the eye of the SPIES rope. It is then connected on the other end of the sling using a Type IV link assembly. The sling must pass between the helicopter skids and the fuselage.

(2) Once the the SPIES rope and cargo straps are in place, the straps running across the deck of the helicopter are secured in place by at least four and as many as eight snaplinks. The snaplinks are evenly spaced across the deck and alternated from one side of the strap to the other and top and bottom, so that the first snaplink is to the rear of the strap and going around the bottom two straps. The next snaplink is in the front of the cargo strap and around the top two sections of the strap. This process continues until at least four points are established.

(3) If eight snaplinks are available, then each tie-down has two reversed. If no hook is available or is not working properly, the SPIES can be used safely by doubling the cargo slings and Type IV links. Thus, two cargo straps are side by side for a total of four slings and four Type IV links. Padding is used around the edge of the fuselage to protect the sling from damage.

(4) Care must be taken when using the UH-1H helicopter because of the ways in which it can be outfitted. Some helicopters have a step attached at the doorway. This is an added obstruction not only during installation but also during the operation. Others may have rocket pods or machine guns mounted. Not all of the UH-1 helicopters are hooked up the same way each time.

4-10. RIGGING OF UH-60 HELICOPTER FOR SPIES OPERATION

The SPIES master inspects all materials required to rig the helicopter.

a. **Equipment.**

- One 120-foot SPIES rope with deployment bag.
- Two 11-foot, 3- or 4-loop cargo slings or two 9-foot, 3- or 4-loop cargo slings.
- Two Type IV connector links (four if aircraft does not have cargo hooks).
- Heavy duty tape (100-mph tape).
- One 12-foot sling rope.
- Five oval snaplinks (nine if the aircraft does not have cargo hooks).

Note: If the aircraft does not have cargo hooks, use four 11-foot or four 9-foot, 3- or 4-loop cargo slings.

b. **Installation.**

(1) The primary attachment point for the SPIES rope is the cargo hook. The end of the SPIES rope has a polyurethane encapsulated eye that is attached to the cargo hook. The two 9- or 11-foot cargo suspension slings are joined to form one continuous sling using a Type IV link. The sling is stretched out on the helicopter deck. One end is taken under the helicopter and through the eye of the SPIES rope. It is then connected on the other end of the sling using a Type IV link assembly. The sling passes between the helicopter skids and the fuselage. Padding is used around the edge of the cargo hatch to protect the sling from damage.

(2) Once the SPIES rope and cargo straps are in place, the straps running across the deck of the helicopter are secured in place by at least four and as many as eight snaplinks. The snaplinks are spaced evenly across the deck and alternated from one side of the strap to the other and top and bottom. Thus, the first snaplink can be to the rear of the strap, wrapping around the bottom two straps. The next snaplink is in the front of the cargo strap and around the top two sections of the strap. This process continues until at least four points are established.

(3) If eight snaplinks are available, each tie-down has two snaplinks connecting the same spot, and the swing gates are reversed. If no hook is available or is not working properly, the SPIES can be used safely by doubling the cargo slings and Type IV links. Two cargo straps are side by side with a total of four slings and four Type IV links.

4-11. RIGGING OF CH-46/CH-47 HELICOPTER FOR SPIES OPERATION

The SPIES master inspects all the materials required to rig the helicopter.

a. **Equipment.**

- Two 11-foot, 3-loop slings.
- Two 9-foot, 3-loop cargo slings.
- Four Type IV connectors.
- One 13-foot sling rope.

b. **Installation.**

(1) The SPIES rope is attached using two 9- or 11-foot cargo suspension slings and four Type IV links. The cargo slings are passed through the encapsulated eye of the SPIES rope and attached to the outboard cargo tie-down rings on the aircraft floor. Two tie-down rings are used for each sling. Padding is used around the edge of the cargo hatch to protect slings from damage.

(2) Not all of the tie-down rings will be in the exact same position on the helicopters. This is one of the main considerations in deviating from the installation procedures. When possible, the cargo straps are placed to form two U shapes. One strap is placed forward of the cargo hole in the center of the aircraft floor and one aft or toward the rear of the helicopter. Installed properly, the cargo straps hold the SPIES rope comfortably in the center of and slightly below the opening of the cargo hatch.

(3) The use of snaplinks attached close to all four faulty tie-down points not only ensures a backup in case of a faulty tie-down ring but also reduces the amount of movement in the cargo suspension straps. A total of eight snaplinks are used for added security at each point with the swing gates reversed.

Section III. LAND AND WATER EXTRACTION

If the situation, mission, or terrain suggests the possibility of a SPIES extraction, the leader includes the SPIES harness in each individual's equipment list. If the mission or insertion precludes the wearing of the harness, it will be carried on the top and just inside the flap of the pack being used. Once the extraction helicopter has been requested, the harness is retrieved and donned.

4-12. LAND EXTRACTION PROCEDURES

The SPIES is used only when the team requires immediate extraction or is unable to move to a clear (open) position suitable for helicopter landing.

a. The extraction helicopter proceeds to the area and establishes radio or visual contact with the team. The backup helicopter equipped with the SPIES remains aloft and away from the area. The helicopter maintains visual contact with the LZ and monitors radio communications.

b. After the team has indicated readiness for pickup and the tactical situation has stabilized, the extraction helicopter moves to the LZ by the safest route. When the helicopter is above the team's location, the SPIES master drops the SPIES rope on order of the pilot after the aircraft has obtained a stable hover at slightly above treetop height.

c. The team hooks up the same as in the familiarization procedures. Team members sling individual weapons over the shoulders, securing weapons and equipment to withstand the wind. A safety line is attached to prevent lost weapons during SPIES operations. When the team is ready, the team leader gives a thumbs-up signal.

d. During extraction, the team RATELO maintains communications with the extraction helicopter. He gives a verbal backup to the thumbs-up signal and relays all information during the flight. His location is near or at the bottom hookup point so he can assist in giving accurate information about the extraction, clearing of obstacles, and descent.

e. Lift-off of the extraction helicopter is vertical until the SPIES rope has cleared all obstacles. With the barrel directed downward at a 45-degree angle and outward, the team members fire their individual weapons using the hip position.

f. Once the SPIES rope is clear of all obstacles, the extraction helicopter changes to horizontal flight and departs the area by the safest route. Airspeed is limited to 70 knots in moderate climates and 50 knots in cold climates while team members are attached to the SPIES rope.

g. When the extraction helicopter has reached a safe dismount area, a transaction is made into hover flight at an altitude of 250 feet AGL. A vertical descent is started with the SPIES master continuously providing information to the pilot on the distance from the ground to the lower end of the SPIES rope. The vertical descent rate of the aircraft (at touchdown) is less than 5 feet per second.

h. When the team reaches the ground, they immediately move out toward the front of the helicopter. All team members rapidly unhook themselves and teammates that need assistance. Once this is completed, they either move away from the area and set up security or assist in clearing the rope if the helicopter is going to land.

4-13. WATER EXTRACTION PROCEDURES

The SPIES is suitable for extracting teams from the water. For this procedure, three inflatable life vests or any type of floatation device is tied to the SPIES rope to provide buoyancy for the rope while in the water. One floatation device is tied at each end of the attachment points, and one floatation device is tied in the middle of the attachment point area just above the middle two sets of D-rings. The team members to be extracted wear their SPIES harnesses under their individual life vest. They can also wear swimming fins, masks, and snorkels (amphibious operations) to ease hooking up to the SPIES rope within the spray area beneath the hovering helicopter.

a. After the extraction helicopter has attained a stable hover above the team member's location, the SPIES master drops the SPIES rope (with floatation attached) on order from the pilot. When the team members complete the hookup to the SPIES rope, the team leader signals the SPIES master to start lift-off. Lift-off of the extraction helicopter is vertical until all team members and the bottom end of the rope have cleared the water. During the initial lift-off, team members know that they are to be dragged through the water. The team members prepare to roll on their backs until clear of the water.

b. Flight speed and altitude are the same as over land. The dismounting procedures remain the same, except when landing on a ship. Once on board, the team members take their orders from personnel in charge of the deck.

Section IV. AFTER-OPERATIONS PROCEDURES

After-operations procedures include repairing, cleaning, and proper storage of the rappelling equipment.

4-14. REPAIRING AND CLEANING OF EQUIPMENT

The contaminated ropes are washed with a mild detergent, such as liquid dish soap, and cold water followed by a rinse in clean, fresh water. Ropes are dried at a temperature not to exceed 140 degrees Fahrenheit. Stubborn oil, grease, hydraulic fluid, and other petroleum stains can be removed with the cleaning agent xylene (Grade A or B, TT-X 916).

4-15. STORAGE OF EQUIPMENT

To avoid ultraviolet deterioration, the nylon materials should be protected from direct sunlight. The SPIES rope is stowed in an aviator's kit bag for protection when not in use. Bins or similar facilities are used for storage of SPIES equipment. Shelves used for storage should be at least 4 inches from the walls and 12 inches from the floor. Areas used for storage should be well ventilated and free of oil, acid, cleaning compounds, and other contaminants. Equipment must *not* be stowed above or near hot water pipes or heating apparatus.

CHAPTER 5
FAST-ROPE INSERTION AND EXTRACTION SYSTEM

The fast-rope insertion and extraction system (FRIES) is employed by small units to make rapid infiltration and exfiltration using rotary-wing aircraft in confined areas. The use of FRIES is normally restricted to situations where aircraft cannot land, and when mission requirements preclude the need for large amounts of equipment, and ammunition or heavy crew-served weapons.

Note: Headquarters, Department of the Army (HQDA) policy specifies that FRIES is not approved for Army-wide use and names the Commanding General, United States Army Special Operations Command (CG USASOC) as the executive agent for FRIES doctrine. The use of FRIES is restricted to special operations forces, pathfinders, long-range surveillance units, and HQDA-approved schools with a USASOC-approved FRIES program of instruction. Requests for other units to conduct FRIES operations should be submitted to Commander, USASOC, ATTN: AOOP-TRS, Fort Bragg, NC 28307-5000.

Section I. GENERAL

FRIES information is subject to change. Units that recognize the need to update this chapter should submit recommendations to Commander, USASOC, ATTN: AOOP-TRS, Fort Bragg, NC 28307-5000.

5-1. OBJECTIVES

The objectives of this chapter are to—

- Prescribe the requirements for FRIES qualification and training, including proficiency sustainment, safety, and administrative procedures.
- Establish the scope of the duties and responsibilities of key personnel who conduct FRIES operations.

5-2. GUIDANCE FOR COMMANDERS

Units that have HQDA approval to perform FRIES operations are authorized to conduct initial FRIES qualification and FRIES master (FRM) qualification training. Training requirements are established in this publication, USASOC Regulation 350-6, and applicable SOAR policies. Aircraft material requirements are specified by the US Army Technology Applications Program Office (TAPO) .

a. FRIES training and operations are inherently risky; therefore, safety is paramount. All training and operations require in-depth attention to detail from planning through preparation and execution.

b. The following basic guidance applies to FRIES training and operations:

(1) Unit commanders must personally approve FRIES training sites.

(2) A risk analysis and or assessment must be completed before FRIES operations.

(3) Night FRIES operations are medium risk or higher.

(4) Units authorized to conduct FRIES operations may incorporate FRIES activities in field training exercises only after personnel have become FRIES qualified.

(5) Field exercises using FRIES activities are subject to the limitations imposed by applicable directives, including mandatory support requirements.

(6) Requests for waiver of procedures or restrictions prescribed herein must be submitted in writing with endorsement of the first general officer in the requesting unit=s chain of command to HQ USASOC.

5-3. TRAINING PREREQUISITES

CAUTION

Trainers must ensure all personnel preparing to participate in FRIES training are tested *with equipment* to ensure they have sufficient upper body strength to safely perform the full scope of roping duties.

Before allowing soldiers to participate in FRIES training and operations, the unit commander must make sure the soldiers are physically fit. Minimum screening standards that apply to each training participant include—

- Having passed the Army physical fitness test within the past six months.
- Having passed a medical examination within the past two years.
- Being free of any injury or physical condition that could cause a potential safety hazard during FRIES training.
- Demonstration of ability to perform a controlled descent from a height of 15 feet while carrying a 40-pound load (not including the basic duty uniform and combat boots).
- Demonstration of ability to hold a static position on a FRIES rope for 20 seconds, using hands and feet to lock in while carrying a 40-pound load.
- Being qualified and current to fly FRIES activities IAW HQ USASOC and the Special Operations Aviation Regiment (SOAR) Aircrew Training Program and applicable policies (for aircrews).

Note: The peacetime maximum soldier load will not exceed 50 pounds. This includes helmet, weapon, vest, web gear, and rucksack. Rucksack weight will not exceed 35 pounds.

5-4. PERSONAL EQUIPMENT REQUIRED

The minimum personal equipment required for FRIES training or operations includes—

- Heavy leather gloves.

- A helmet with a chin strap.
- Protective goggles.
- A long-sleeve shirt or jacket, long pants, and boots.
- Hearing protection and identification tags for helicopter operations.

Section II. FRIES QUALIFICATION TRAINING

Each major unit (commanded by an O5 or higher) is responsible for conducting its own FRIES qualification and sustainment program. FRIES trainers will be FRIES master (FRM) qualified and current. Unit training should be tailored to unit and situation needs, however proponent requisites will not be reduced without written approval of the CG USASOC.

5-5. INITIAL FRIES QUALIFICATION TRAINING

Before participating in fast-rope operations, personnel are briefed on the FRIES and its purpose, capabilities, and limitations. The briefing also covers the duties and responsibilities of the pilot in command (PIC), safety, FRIES master (FRM), the assistant FRIES master (AFRM), and any ground assistants. Once the FRIES briefing is conducted, the remainder of the initial training is hands-on practice of the proper FRIES operational techniques.

a. Individuals are shown the proper techniques for boarding the aircraft, moving to the door, grasping and descending, locking-in, and clearing the rope. Individuals are also shown the proper wear, hook-up, and use of the STABO extraction harness.

b. After the demonstration, all ropers participate in a practice exercise in which they properly perform the following tasks using a tower and subsequently an aircraft. All ropers must demonstrate a minimum of six properly executed FRIES descents (three without equipment and three with equipment) and at least one successful lock-in.

(1) *Perform FRIES descent and lock-in from a tower at the 34-foot level.* Training should be progressive, starting from a tower at the 34-foot level without equipment and then with equipment. All ropers must complete a successful lock-in at this level. Ropers will not progress above the 34-foot level until after demonstrating the ability to stop descent, lock-in, and hold a stationary position for 20 seconds with equipment.

(2) *Perform FRIES descent and lock-in from a tower at the 60-foot level.* The recommended succession will be no less than three successful FRIES descents on the 60-foot tower, again starting without equipment, completing a lock-in, and finishing with equipment.

(3) *Perform FRIES descent from a helicopter at 60 feet or higher.* The recommended completion should be at least two helicopter descents from 60 feet or higher (one without equipment and one with equipment).

WARNING
During initial qualification, students will not perform rapid exits or have more than three soldiers on a rope at any time. During advanced training, the limit is no more than five personnel with equipment on any rope at any time. In all cases (insertion or extraction), the maximum load allowed on any fast-rope system will not exceed 1,500 pounds.

CAUTION
All ropers should be limited to no more than a total of 10 roping events in a 24-hour period; no more than 6 events should be with equipment.

5-6. FRIES PROFICIENCY SUSTAINMENT TRAINING

Commanders must make sure soldiers participating in FRIES operations receive sustainment training on equipment and procedures within 24 hours before the FRIES operation. Ropers who do not attend FRIES sustainment training will not be allowed to participate in FRIES operations. As a minimum, training will include a review of the following:

- Arm-and-hand signals.
- Individual equipment riggings.
- Aircraft familiarization.
- Safety procedures.
- Any rehearsals the FRM or commander deems necessary.

Section III. FRIES MASTER SELECTION AND QUALIFICATION TRAINING

Selection and qualification of FRIES masters (FRMs) is a unit prerogative. Units will maintain records of qualification and proficiency. These records will accompany SM to new units when a permanent change of station occurs.

5-7. FRIES MASTER SELECTION

Units may train and certify soldiers as FRIES-qualified FRMs. Soldiers selected by units to qualify as FRMs should be the rank of E5 or higher and should have demonstrated leadership abilities with emphasis on maturity and decisiveness.

5-8. FRIES MASTER PREREQUISITES

Soldiers performing FRM, AFRM, and safety duties must have completed initial FRIES qualification training, be experienced in FRIES operations, have completed FRM training, and have executed at least 10 FRIES descents and 2 extractions. An FRM candidate's experience must include no less than five descents with combat load and at least one night descent with equipment from 60 feet or higher.

5-9. FRIES MASTER TRAINING AND CERTIFICATION

During FRM training and certification, FRM candidates must participate in three FRIES helicopter operations (twice observing another FRM and once executing FRM duties under observation of a current FRM) and be certified by either the parent unit or an authorized school. FRMs must be proficient in FRIES operations and must demonstrate proficiency at the following:

- Inspecting, preparing, and rigging all FRIES gear.
- Inspecting and preparing aircraft for FRIES operations.
- Controlling and coordinating actions of AFRMs, safeties, and ropers.
- Preparing and conducting both insertion and extraction operations.
- Conducting troop and pilot briefings.
- Giving arm-and-hand signals.
- Giving time warnings and associated commands.
- Deploying and retrieving FRIES ropes.
- Rigging and lowering equipment.

5-10. FRIES MASTER REFRESHER TRAINING

Units will conduct refresher training to maintain acquired skills. FRMs who have not participated in FRIES operations during the past six months will receive refresher training by a current FRM and serve as an AFRM before performing FRM duties. Refresher training for ropers consists of an FRM briefing and participation in FRIES training.

Section IV. KEY PERSONNEL DUTIES AND RESPONSIBILITIES

The following personnel duties and responsibilities provide baseline requirements for the safe conduct of FRIES operations. Unit standing operating procedures (SOPs) may increase (but will not reduce) training safety requirements.

5-11. AIR MISSION COMMANDER

The air mission commander (AMC) provides overall command and control when conducting large, complex missions. The AMC is selected or appointed by the headquarters responsible for the overall mission. He ensures that all operators, aircrew members, and support elements synchronize their actions during the conduct of the mission and the FRIES operation. An AMC is not usually needed during small-unit qualification or proficiency training events when only one helicopter is used.

5-12. FRIES TRAINING OFFICER

Units that conduct FRIES training appoint a training officer in charge (OIC) or a noncommissioned officer in charge (NCOIC) to manage the training. The OIC is responsible for planning and coordinating various aspects of training to include—

- Coordinating all support activities, such as procuring FRIES equipment, aircraft, training areas, medical support, and communication.

- Assigning qualified personnel to perform FRM, AFRM, ground safety, and medic duties.
- Adhering to procedures for planning, preparing, and executing the operation IAW this TC, USASOC Regulation 350-6, and the participating unit=s SOPs.
- Ensuring the FRMs, AFRMs, PICs, aircrews, safeties, and medics are briefed on the operation.
- Ensuring FRIES training events are not conducted on or over unsafe ground. (The ground must be free of loose material that could become airborne and cause an unsafe condition or obscure vision for aviators or ropers.)
- Ensuring a medic, an aid bag, a backboard, and a dedicated vehicle and driver are on site during all training.
- Ensuring positive communication with aircraft is maintained throughout training.

5-13. PILOT IN COMMAND

The aviation unit providing helicopter support for the FRIES training and operation appoints the PIC. The PIC is responsible for all aspects of the flight and ensures—

- Aircrew members are current and qualified to conduct FRIES operations
- Aircrew members are informed of and understand their responsibilities in fast-rope operations.
- Procedures for planning, preparation and execution are adhered to IAW this TC, USASOC Regulation 350-6, and applicable SOPs and policies.
- Aircraft is properly configured to perform the mission.
- Aircraft is at the proper altitude, airspeed, and location as briefed.
- The command to deploy ropes is not given until the aircraft is at a stabilized hover.
- The aircraft position is maintained to keep ropes in contact with the surface until all descending ropers are on the ground (or extracting ropers are securely attached for extraction).
- Ropes are fully recovered inside the aircraft (or jettisoned) before the aircraft departs the stabilized hover position at the infiltration site.
- All personnel are briefed on in-flight emergency and safety procedures.
- Ropes are never deployed with anything other than night illumination attached to the free end.
- All personnel are briefed on fire support to be provided by the aircraft including:
  - — The nature of fire support the aircraft can provide.
  - — The time the fire support starts, shifts, and stops.
  - — The primary and alternate commands and signals that start, stop, lift, or shift fires.

5-14. FRIES MASTER

Units conducting FRIES operations will designate one overall FRM to organize, coordinate, and supervise the activities of the day, and AFRMs as needed. AFRMs are additional FRM-qualified soldiers who assist the primary FRM. An AFRM will be designated for each roping point. AFRMs are responsible for all preparation, inspection, and command and control of all roping activities on their points. FRMs also must—

- Ensure all FRIES roping personnel, including FRMs, AFRMs, and safeties, are qualified and current and understand the proper roping procedures for FRIES operations.
- Perform safety and serviceability checks on all FRIES equipment and rigging.
- Ensure the attachment bar or points are serviceable and free of any defects or contamination and that the quick-release mechanisms and safety pins are present and serviceable and operate correctly.
- Inspect the rope to make sure it has no contamination, damage, or defects that could make it unsafe.
- Check the rope to make sure it is the correct type and length for the operation (smooth rope for infiltration or looped for exfiltration).
- Ensure the rope is properly attached with safety pin in place and back coiled.
- Ensure the extraction harnesses are serviceable and properly worn.
- Ensure the rope chemlights are correctly rigged and illuminated when needed (two at the mount, two at the end, and two 15 feet from the end).
- Confirm the proper seating arrangement, including seat belts or a safety strap for ropers.
- Coordinate with the PIC and brief the aircrew, safeties, and ropers on the following:
  - — The aircraft flight, approach, flare, and position over the target site.
  - — The height of the aircraft above the target.
  - — A description of the FRIES site (rooftop, small clearing) and size, as well as the actions of the ropers upon arrival at the site.
  - — The signal to deploy the rope, who will deploy it, and how it will be done.
  - — Time warnings and actions at each time warning.
  - — Arm-and-hand signals.
  - — The use of radios and intercoms.
- Brief the PIC and crew chief on verbal commands and final adjustments over the target.
- Confirm the correct target and inform the pilot before the deployment of ropes or ropers.

WARNING
The FRM must ensure that nothing, other than rope illumination, is ever attached to the free end of the FRIES rope during deployment and that equipment is never attached and dropped.

- Deploy the ropes when the PIC gives the command DEPLOY ROPES. (This action may be delegated to the aircrew or safety.)
- Ensure safeties and AFRMS understand the time and activity sequence and all arm-and-hand signals.
- Ensure safeties know the proper procedures to—
- Keep the aircraft on target.
  - — Clear the ropes.
  - — Recover the ropes.
  - — Jettison the ropes.
- Observe the target and pass to the PIC, through the crew chief, final aircraft position adjustments over the target to make sure conditions are safe.
- Deploy the ropes when the PIC gives the command DEPLOY ROPES. (This action may be delegated to the safety.)
- Ensure all ropers know how to hook up for extraction and know the correct arm-and-hand signals.
- Direct the aircrew or safety to inform the pilot when the aircraft is on target (the PIC replies with the command DEPLOY ROPES).

CAUTION
When equipment is to be lowered, it should be lowered before the FRIES is deployed or personnel begin roping. When simultaneous, equipment and ropers will exit from opposite doors of the aircraft.

— The safety repeats, " DEPLOY ROPES." The FRM signals the AFRMs to deploy ropes and makes a final check, ensuring all ropes from his aircraft reach the surface and are safe for descent (at least 15 feet of rope must be on the surface).

— When conditions are safe, the FRM commands "GO" to the ropers and points to the rope or leads the stick out.

Note: If the FRM leads the stick out, the AFRM automatically assumes control. He completes marshaling the remaining ropers and is the last man to exit.

- Coordinate actions with the crew chief for the positioning or storage of seat belts, safety straps, rucksacks, door bundles, and communication cords.

Note: During night operations, upon completing descent, the AFRM observes the the ropes of his aircraft. When he sees that all ropes are clear, he issues the prearranged ALL CLEAR hand signal to the aircrew safety. This hand signal tells the aircrew that they are free to jettison or recover the ropes, and fly away.

5-15. FRIES AIRCREW MEMBER AND SAFETY

FRIES safeties are aircrew members who are FRM-qualified and perform the additional duties of the safety during helicopter roping operations. Safeties assist FRMs and pilots as needed. Aircrewmen performing safety duties differ from FRMs and AFRMs in that safeties are members of the aircrew (crew chief or gunner) whose orientation is the flying portion of the operation. Conversely, AFRMs are FRM-qualified members of the ground unit who are performing FRM duties as assistants to the primary FRM and who exit with the ground unit. Safeties remain in the aircraft and operate the doors, extend and retract the rope mount bars, and wear night vision goggles (NVGs) when appropriate. Safeties must have completed FRM training IAW this TC and the aircrew training program (ATP).

a. During roping operations, the safety's four major areas of responsibility are:

(1) Observe all activities in the aircraft and assist the pilot and FRM or AFRM as needed.

(2) Serve as the communication link between the PIC and the FRM. The safety monitors aircraft communication and relays information between the PIC and the FRM, constantly keeping both informed.

(3) Assist the FRMs and AFRMs in lowering cargo and clearing the aircraft of ropers.

(4) Clear and jettison the ropes, when appropriate.

b. The safety performs the following duties:

(1) Relays time warnings, aircraft positioning commands, and other information to assist the FRM.

(2) Conducts aircraft equipment rigging and serviceability inspections, making sure mounts, doors, and ropes are ready and safe for operations both pre-flight and en route.

(3) Verifies that the aircraft is at a stabilized hover with ropes in contact with the surface and remains so (or stops stick until corrections are made).

(4) Checks the rope to make sure 15 feet of rope are on the target surface.

(5) Tells the PIC, "ROPES DEPLOYED", when appropriate.

(6) Tells the PIC, "ROPERS OUT", when the first roper has exited.

(7) Helps the AFRM in controlling and spacing each roper's exit.

(8) Monitors the rope and ropers to make sure they do not get hung up, dragged, or dropped off site.

(9) Observes the exit of the last roper until the roper reaches the surface, moves clear of the rope, and gives the ALL CLEAR signal.

(10) Tells the PIC, "ROPERS DOWN", (station name: aft, forward, left, right) when all ropers are on the surface and conditions are safe to jettison the ropes.

(11) Passes the signals as required to position the aircraft, including HOLD, MOVE LEFT, RIGHT, FORWARD, BACK, UP, or DOWN.

(12) Acts as belay man during cargo lowering when needed.

(13) Jettisons ropes upon receiving the command JETTISON ROPES from the PIC. Tells the PIC, "Ropes are jettisoned," when the ropes have been recovered or jettisoned.

(14) Tells the PIC when ALL CLEAR.

CAUTION
This command tells the pilot the aircraft is cleared for flight.

(15) Deploys the rope for extraction upon receiving the PIC command DEPLOY ROPES.

(16) Observes the extracting rope and personnel and informs or guides the PIC through aircraft positioning and throughout the recovery flight.

5-16. FRIES ROPER

Ropers are responsible for notifying the FRM, AFRM, safety, or pilot if they observe any unsafe act or condition. Ropers may halt or call for a halt of roping for safety at any time. During FRIES insertion training, the number of ropers on the fast rope at one time is limited to three. During extraction training, the total weight per extraction bar or rope will not exceed a total of 1,500 pounds. When rucksacks are worn by ropers, the rucksacks must not exceed a weight of 35 pounds (total roper equipment weight will not exceed 50 pounds). Ropers will—

- Keep hands at head level.
- Maintain visual contact with lower ropers during their descent.
- Maintain a minimum of a 1-second interval on exit to avoid collisions.
- Keep at least two points of contact on the rope (both hands) at all times.
- Use their feet for additional breaking any time needed.
- Execute descents at a safe speed.
- Slow the rate of descent halfway down the rope to avoid landing on each other.
- Move quickly away from the rope upon arrival on the surface.
- Know correct wear of extraction (STABO) harness and extraction procedures.

WARNING
The FRM, AFRM, and safety must make sure all personnel are on the surface and have moved clear of the ropes before aircraft fly away or ropes are jettisoned.

Section V. FRIES RIGGING OF AIRCRAFT

The aviation unit is responsible for preattachment inspection and installation of FRIES anchor point mount bars and associated equipment. All aircraft used to conduct FRIES operations with US Army personnel must be rigged and equipped IAW current parent service directives--for Army aircraft, USASOC, SOAR, and TAPO directives apply. Rigging of the aircraft includes installation and removal of the seats and seat belts (or personnel restraints), and attaching and detaching the ropes.

Figure 5-1. UH-60 helicopter

> **CAUTION**
> **Army aircraft that have the FRIES mounts installed must have an airworthiness release for FRIES operations from the Aviation Troop Support Command (ATCOM) TAPO.**

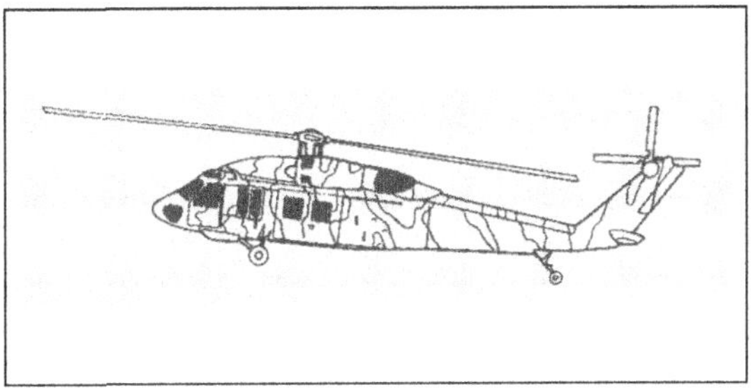

5-17. RIGGING OF FRIES IN UH/MH-60

The UH-60 is a twin-turbine, medium-speed, single-main-rotor helicopter (Figure 5-1).

The aviation aircrew or safety is responsible for—

- Inspecting the FRIES mount bar assembly, attachment hardware, and associated equipment to make sure they are serviceable and ready to use.
- Ensuring the bar assembly is correctly mounted to the aircraft (Figure 5-2).

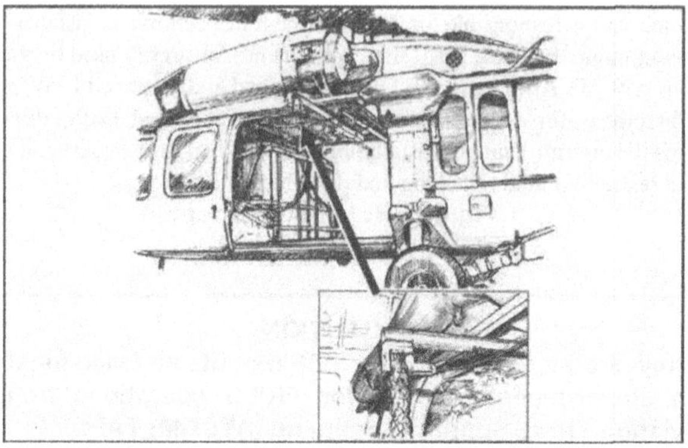

Figure 5-2. Bar assembly correctly mounted to aircraft.

- Ensuring the mount is complete, functional, and free of any defect.
- Ensuring the mount and the aircraft floor are clean and free of any petroleum, oils, or lubricants (POL) or other contaminants that could weaken the rope or cause personnel to slip or fall.
- Ensuring the mount and aircraft are free of sharp edges or burrs that could cut the rope.
- Ensuring no metal fatigue or structural weakness exists that could cause the system to fail.
- Extending locking and retracting mount bars as needed.
- Lowering and belaying equipment as needed.
- Opening, locking, and closing cargo doors as needed.

Note: For cold-weather operations or during flights of long duration, the cargo doors are closed and latched until the time specified for opening. The aircrew will operate doors or directly supervises those who do. In good weather, doors should remain locked open.

- Removing or reconfiguring troop seats.
- Providing seat belts or floor-mounted personnel restraints for ropers' use during flight.
- Attaching and detaching ropes as needed.

Note: Clusters of rucksacks or small door bundles (weight limited to 250 pounds) may be lowered (belayed) on rappelling ropes. Cargo lowered this way must never be free dropped (the belay must start at the time the bundle is pushed [tipped] out the door).

5-18. RIGGING OF FRIES IN CH/MH-47

The CH-47 is a tandem-rotor, medium-transport helicopter (Figure 5-3).

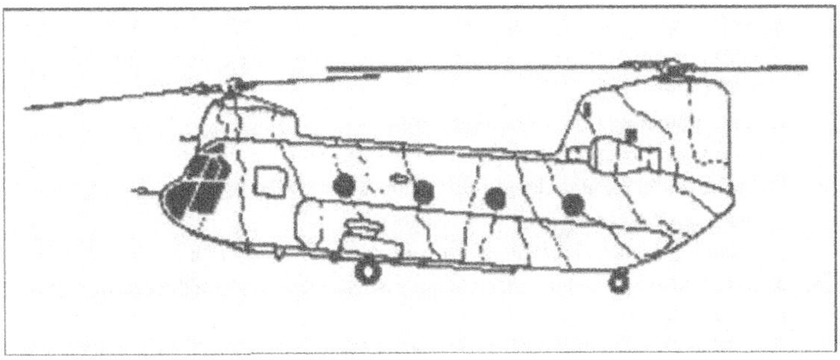

Figure 5-3. CH-47 helicopter.

The aviation aircrew or safety is responsible for:

- Inspecting the FRIES mount bar assembly, attachment hardware, and associated equipment to make sure they are serviceable and ready to use.
- Ensuring the bar mounts are correctly mounted to the aircraft (Figure 5-4, rear; Figure 5-5, winch).

Figure 5-4. Bar mounts correctly mounted, rear.

Figure 5-5. Bar mounts correctly mounted, winch.

- Ensuring the mount is complete, functional, and free of any defect.

- Ensuring the mount and the aircraft floor are clean and free of any POL or other contaminants that could weaken the rope or cause personnel to slip and fall.
- Ensuring the mount and aircraft are free of sharp edges or burrs that could cut the rope.
- Ensuring no metal fatigue or structural weakness exists that could cause the system to fail.
- Extending locking and retracting mount bars as needed.
- Lowering and belaying equipment as needed.
- Opening, locking, and closing cabin doors as needed.
- Removing or reconfiguring troop seats.
- Providing personnel restraints for ropers' use during flight.
- Taping and padding any sharp edges that can come in contact with ropes or personnel
- Attaching and detaching ropes as needed.

5-19. RIGGING OF FRIES IN MH-53

The MH-53 is a USASOC twin-engine, single-rotor, medium-transport helicopter used by the Army (Figure 5-6).

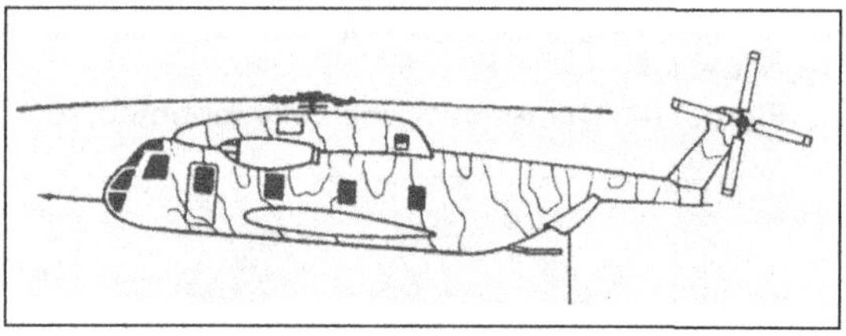

Figure 5-6. MH-53 helicopter.

The MH-53 primary FRIES mount bar assembly is located overhead, just aft of the rear ramp. The ramp mount bar assembly spans the cabin width and has a two-pin release mechanism on each end. The ramp position mount assembly provides two points; each will accommodate one FRIES rope or one cargo lowering rope. The ropes are carried back-coiled on the ramp.

a. The alternate position for fast-rope descent from the MH-53 is the crew entrance in the right side of the fuselage immediately behind the pilot's position. In this position, the fast-rope mount must be connected to the external hoist point. The FRIES master is responsible for inspecting the MH-53 mount system for its presence, security, and operability.

b. The aviation unit is responsible for—

- Installation and maintenance of the fast-rope mount assemblies at each

position to be used.
- Configuring the aircraft to accommodate the number of soldiers and type of mission planned.
- Ensuring seat belts or restraint straps are present to secure ropers during flight.

Note: The MH-53 aircrew can improvise handrails to assist roper exits. When used, improvised handrails will be rigged IAW USAF special operations SOPs.

- Configuring the MH-53 for fast-roping when equipped with external 650-gallon fuel tanks.
  - Remove the ramp door to install the cross bar.
  - Place the bar in the brackets with the two quick-release paddles facing forward.
- Configuring aircraft armament (50-caliber machine gun and the GAU-2B ramp minigun) for FRIES operations.
  - When using the aft right roping position, point the 50-caliber machine gun to the 6 o'clock position and secure the gun with bungee cords.
  - When using the left aft rope, disconnect the minigun ammunition chute and stow it in the ammunition can, or disconnect it at the ammunition can and securely wrap it around the minigun. Use bungee cords to secure the ramp minigun in the firing position facing the 6 o'clock position.

Note: Ideally, only the right aft rope is used when the GAU-2B ramp minigun is installed. Although not recommended, the left aft rope may be used, if required.

Section VI. EQUIPMENT MAINTENANCE AND INSPECTION

Before conducting a FRIES operation, the FRM must inspect all FRIES equipment (ropes, STABO harnesses, and mount bars) for serviceability and readiness for use.

> **WARNING**
>
> **FRIES ropes and harnesses are critical life support equipment and must be stored, cared for, and maintained as life support items.**

5-20. ROPES AND HARNESSES

FRIES equipment should be maintained in the same manner as a parachute. The unit rigger section can provide detailed guidance on appropriate inspection, care, and maintenance of FRIES equipment.

a. FRIES ropes and harnesses must always be stored in a clean, cool, dry space out of direct sunlight and free of chemicals or chemical vapors. Equipment that becomes wet with fresh water should be hung up to dry (indoors) on hardwood pegs. Equipment that is exposed to salt water or becomes imbedded with dirt or mud should be washed and rinsed in fresh water (within 72 hours) and hung up to dry (indoors) on hardwood pegs (out of direct sunlight).

b. Before conducting a fast-rope operation, the FRM—

- Inspects the fast rope thoroughly and carefully. Checks the rope length to ensure it is the correct rope for the operation planned.
- Checks the woven loop on the mount end for excessive wear or chemical contamination. Checks the rope along its entire length for fraying, cuts, and chemical contamination.
  - Do not use a rope that is severely frayed. (Light fraying on the rope from normal use does not weaken the rope.)
  - Do not use a rope when any single strand is cut halfway through or has two or more cuts that penetrate one-third or more through any strand's thickness within 1 foot of the running length of the FRIES.
- Inspects the rope for contamination of acid, alkaline compounds, saltwater, fire extinguisher solutions, or petroleum-based solvents. Changes in color caused by chemicals are usually blotchy and have an unusual odor. Although used ropes gradually change color, such changes do not indicate a decrease in strength unless the change is due to contact with strong chemicals. Changes occurring because of use are usually uniform throughout the length of the rope.

WARNING

Any rope known or suspected to be chemically contaminated, cut beyond allowable limits, excessively frayed, or sun damaged must be taken out of service, marked for destruction, and turned in to the rigger section.

- Inspects the extraction loops to the same standard as the main rope. Ensures the woven attachment loops are secure.
- Inspects the harness to ensure—
  - Ropers are wearing the harness under all load-carrying equipment.
  - Ropers have properly fastened all connectors.
  - Harness material and stitching are not cut, torn, or contaminated, and all hardware is free of corrosion and is in operable condition.

Note: Serviceability inspection and maintenance of the extraction harness are the responsibility of the rigger-qualified personnel.

5-21. FRIES MOUNT BARS

The aviation unit is responsible for the installation, removal, storage, and maintenance of FRIES mount bars. The US Army Special Operations Aviation Regiment, Ft Campbell, KY, prepares and provides checklists for these procedures. Aviation units participating in FRIES operations must have and comply with current SOAR guidance.

Aviation units:

- Install and maintain the FRIES mount bar; check mount bolt torque and witness marks.
- Check the FRIES mount bar for security and operability.
- Check the condition of the mount bar; inspect for bends, dents, distortion, cracks, corrosion, and contaminants.
- Ensure the fast-rope quick-release pins are installed and operable (two each side).
- Check the security and condition of the rope-release system, as follows:
  - The mounting bolts are installed and properly torqued.
  - The safety pins are installed, and lanyards are present.
  - The system locks and unlocks as designed.
- Ensure the release cables are attached at both ends and are not broken or kinked.
- Check the release handles to ensure they are securely mounted and can move freely.
- Check the latch as follows:
  - Latch or close the release mechanism and apply a downward pressure on the release arm. Attempt to pull the release handles the release should not open.
  - Relax pressure on the release mechanism and pull the handles the release should open.
  - With no pressure on the release mechanism and with the safety pin installed, attempt to pull the release handles--the release should not open.

Section VII. OPERATIONAL REQUIREMENTS AND LIMITATIONS

This section establishes medical coverage and communication support requirements for FRIES operations, including during adverse weather conditions and night operations.

5-22. MEDICAL SUPPORT

Dedicated medical support is required for FRIES training and demonstrations. A qualified medic with assistant, aid bags, two backboards, two stretchers, and a dedicated vehicle with a driver must be present to respond to any mishaps. The nearest medical evacuation (MEDEVAC) unit will be notified of where and when FRIES training is scheduled. MEDEVAC response will be coordinated as on call. Training or demonstrations will not be conducted without medical support present. The ground force commander may direct the FRIES aircraft to transport life or death (urgent) patients when no other MEDEVAC unit is available, providing the medical team is trained and properly equipped with the appropriate equipment for aerial transport of patients. This must be included in the air mission briefing prior to training and coordinated with the medical team providing coverage. FRIES operations that are evaluated as high risk or that occur at night must have a certified combat lifesaver on location and medics (equipped as stated above) in the immediate vicinity.

5-23. COMMUNICATIONS REQUIREMENTS

During peacetime training and operations, FRIES lift aircraft and ground elements will maintain continuous radio communications. Aircrew personnel clearing ropes at each FRIES station must be on the intercom system, monitoring aircraft internal communications. They must also be proficient in the use of arm-and-hand signals as precoordinated between FRIES users and aircraft crews. The designated aircrew member will relay time warnings between the PIC and FRM. Routine time warnings are 10-, 6- and 1-minute warnings to the FRM. Ground-to-air communications must be continuously maintained throughout FRIES training.

5-24. ADVERSE WEATHER OR TERRAIN CONDITIONS

FRIES training and operations will ***not*** be conducted under the following conditions:

- Anytime temperatures are below 32 degrees Fahrenheit or wind-chill factors that could cause cold-weather injuries exist including those caused by helicopter rotor wash or extraction cruise speed.
- Anytime water or ice on the rope prevents personnel from controlling their descent.
- Anytime the rope is exposed to freezing conditions for a sufficient length of time to freeze, thereby reducing its tensile strength.
- Anytime dust out or white out conditions restrict the aviators or the ropers vision.
- Anytime trees or vegetation on the LZ is too dense for safe training.
- Anytime the LZ is too steep or too slippery for safe training.
- Anytime winds exceed 30 knots or are too gusty for safe operations.

5-25. NIGHT OPERATION REQUIREMENTS

Night operations require the following procedures.

a. Attach two chemlights at the free end of the rope and two above the extraction attachment loops (15 feet from the end of the rope) to aid in determining when adequate rope is on the surface.

b. Attach two chemlights (one primary and one backup) to the loop of the rope, or to the rope mount, to guide personnel exiting the aircraft to prevent glare from obscuring vision.

c. Attach one chemlight or luminous tape to the top of each roper's helmet. If a chemlight is used, select a color different from the colors used on the ropes.

CAUTION
Ropers will not wear NVGs during descent. NVGs limit depth

perception and cause a tunnel-vision effect. Static FRM and Safeties may wear NVGs to help confirm that fast ropes are fully deployed. The aircrew will wear NVGs as required by flight mission needs.

Section VIII. FRIES PROCEDURES

The risks of personal injury and equipment damage during FRIES operations can be reduced by detailed planning, thorough training, and strict adherence to safe procedures. Conducting FRIES operations with heavy loads requires personnel to be proficient in these operations. FRIES users must limit ropers to no more than 6 descents with combat load and no more than 10 total descents in any 24-hour period. The total weight of equipment soldiers may carry during training is limited to never more than 50 pounds. This section provides standardized procedures for FRMs, PICs, and ropers. These procedures are common to all units and aircraft, except as noted.

5-26. FRIES MASTER

The FRM performs the following series of actions.

- Briefs members of his team and aircrew.
- Inspects team members and equipment.
- Installs the FRIES rope in the aircraft and conducts safety checks.
- Relays time warnings to team members. (Time warnings are a tool to help keep aircrew and ropers' actions synchronized and can be modified according to user needs; however the 1-minute warning should always be used.)
- Breaks chemlights, if required. Two chemlights or luminous tape will be attached at the anchor point to identify the handhold point.

WARNING

Equipment must never be attached to the FRIES and dropped. As little as 10 pounds of weight can exceed the load limit of the FRIES mounts when the load stops at the end of a 60-foot drop. The only items to be attached to the fast rope during deployment are the position markers.

- Ensures the rope is properly configured for deployment (back coiled to prevent tangles).
- Ensures the team members are in order of exit no later than the 1-minute warning.
- Confirms target on final approach.
- Deploys the rope and ensures it reaches the ground. During night

operations, the rope is marked with six chemlights: two at the mount, two on the end, and two 15 feet from the end.

- Deploys personnel.
- Accounts for personnel, and signals aircrew.

WARNING

Because the FRIES mount in the CH-47 helicopter is narrow, special care must be taken by ropers to avoid fouling weapons or equipment on the frame during exit.

5-27. PILOT IN COMMAND

The PIC briefs the FRM, safety, and roping personnel before FRIES operations.

a. The PIC briefing includes:

- Approaching, loading, unloading, and departing the aircraft.
- Actions in the aircraft.
- Flight route and checkpoints en route.
- Altitude or height for roping.
- Time warnings to FRM as needed (recommend 10 minutes, 6 minutes, and 1 minute).

b. When the PIC is over the intended objective he calls, "DEPLOY ROPES" Fast-rope warnings and expected aircrew coordination calls are as follows:

(1) The aircrew or FRM provides any final position adjustments of the aircraft with the commands of MOVE (left, right, forward, back, up, down, or hold) as needed to place the ropes ON TARGET. (When the pilot is to hold the position, an explanation will be provided as soon as possible.)

(2) When the FRM deploys the ropes, the aircrew member calls, "ROPES DEPLOYED."

(3) When the first roper exits the aircraft, the aircrew member calls, "ROPERS OUT."

(4) The safety watches the last roper until he is on the ground and gives the prearranged signal. The safety then tells the PIC, "ALL ROPERS AWAY," and identifies his station. This alerts the PIC that the safety is going to either jettison or recover the ropes.

(5) The safety will then either recover or jettison the rope as planned, clear the aircraft all around for obstructions, and call, "ROPES CLEAR (station)" This informs the PIC that the aircraft is clear for flight.

c. The PIC also briefs on emergencies that could occur in the following situations:

- En route (when ropers are inside the aircraft).
- Before deployment of ropes.

- When fast-rope personnel are on the ropes.
- After the ropers are on the surface.

5-28. ROPERS

Ropers perform the following actions:

a. At the command STAND BY (given at 1-minute warning), ropers make final check of themselves and take the prepared-to-exit position.

b. At the command GO, the first roper exits the aircraft, rotates his body 90 to 180 degrees to make sure his equipment clears the aircraft, places the fast rope between the arches of his feet, and begins descent. The roper uses his hands and feet to slow his descent approximately two-thirds of the way down.

c. Subsequent ropers exit at 1-second intervals using the same procedure, but begin slow descent approximately half-way down to avoid landing on other ropers.

d. During descent, ropers keep a lookout and break as necessary to avoid landing on obstructions or on fellow ropers.

e. Ropers prepare to land just before reaching the ground by spreading their legs about shoulder-width apart with their knees slightly bent.

f. At landing, ropers quickly move clear of the ropes to avoid collisions with descending ropers.

CAUTION

According to AR 95-3, anytime operations occur over water, all personnel will wear an approved personal flotation device as identified in FM 57-220 or USASOC Regulation 350-2.

5-29. EQUIPMENT-LOWERING PROCEDURES

Anytime a roper's load exceeds a total of 50 pounds or his rucksack weight exceeds 35 pounds, equipment will be lowered (as clusters or bundles) on separate ropes. Lowering equipment from heights great enough to cause equipment damage necessitates belaying the equipment during descent to avert damage. Belaying devices used in mountaineering allow ropers or safeties to control equipment descent.

Figure 5-7. Rucksacks upright on cargo deck.

CAUTION

When equipment is to be lowered, it should be lowered before the FRIES is deployed or personnel begin roping. When done

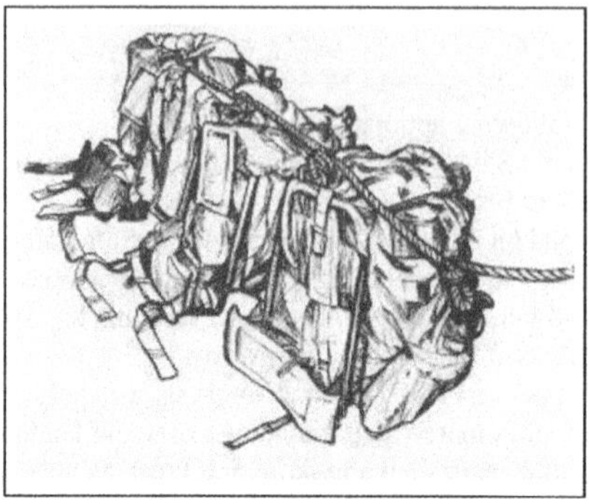

simultaneously, equipment and ropers will exit from opposite doors of the aircraft.

a. Rigging equipment to be lowered requires rigging rucksacks in clusters. Clusters of up to five rucksacks may be lowered (smaller clusters work better). FRMs will ensure the total weight of any cluster does not exceed 250 pounds.

(1) Place the rucksacks upright on the cargo deck of the helicopter (Figure 5-7) .

(2) Lay a sling rope across the frames between the shoulder strap attachment points.

(3) Tie a bowline knot at each end of the sling and a figure eight knot at each rucksack.

(4) Attach a snaplink to the frame of each rucksack, to the loops in the sling rope, and to the end of the lowering rope.

(5) For Lowe rucks, install the issue H-harness on the rucks and join them into clusters with a sling rope tied to the quick-release attachment buckles.

b. Rigging of the lowering rope to belay the rucksacks requires the following actions.

(1) Backcoil a rappel rope on the helicopter floor at the gunner=s position.

(2) Attach a snaplink and belay device (either a figure eight [Figure 5-8] or an SMC ladder [Figure 5-9]) to the floor tiedown (with a snaplink) inboard of the ceiling rappelling anchor point.

(3) Attach a snaplink to the ceiling rappelling ring.

(4) Route the belay rope from the backcoil, through the belay device, then up through the ceiling snaplink, and down to the cluster snaplink.

(5) Tie the end of the belay rope into a bowline with halfhitch and snap it into the cluster snaplink.

(6) The FRM identifies two ropers to push the cluster or bundle out of the aircraft. He also identifies an AFRM or a safety to belay the cluster or bundle. The FRM briefs these personnel on the time to lower the bundle, the signals to be used, and the procedure to clear the belay system before deploying the FRIES. In performing his

belaying duties, the AFRM or safety must The aviation aircrew or safety is responsible for:

- Provide sufficient slack to allow the pushers to eject the cluster from the aircraft.
- Control the descent until the cluster reaches the surface.
- When the cluster is on the surface, unsnap both the upper and lower snaplinks, drop the remainder of the rope from the aircraft, and immediately look out to make sure conditions are safe to deploy the FRIES.
- Inform the FRM and pilot when conditions are safe.

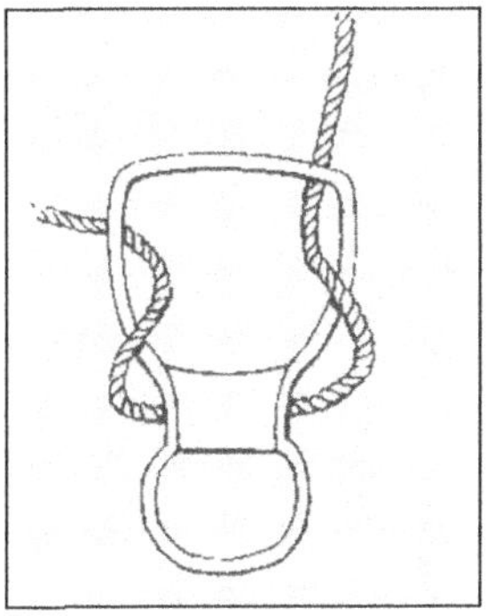

Figure 5-8. Belay Device figure eight SMC

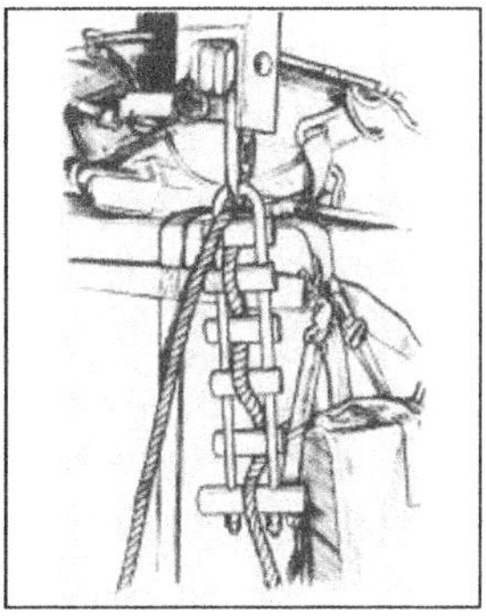

Figure 5-9. Belay device ladder

- Deploy FRIES on command

Note: Expect to reposition aircraft before ropers exit to avoid ropers landing on equipment.

5-30. HELICOPTER OPERATIONS FRM CHECKLIST

Table 5-1 shows a checklist of duties of key personnel in fast-rope operations.

PRE-ROPING ACTIONS
Receive a briefing from the OIC or the AMC.
Coordinate and brief all participants.
Rig aircraft and conduct a joint inspection.
Brief ropers, AFRMs, and safeties.
Rig and inspect ropers.

Conduct a static rehearsal.
AIRCRAFT LOADING Position equipment and personnel. Make sure all personnel have straps or seat belts. **ACTIONS IN FLIGHT** Monitor the command net. Monitor the aircrew net. Monitor the flight route. **ACTIONS AT 10-MINUTE WARNING** (applies to long infiltrations) Issue the 10-minute time warning and GET READY. Check equipment and belay system hookup. Check fast-rope hookup. Secure fast-rope bar in position. Make sure rope is back coiled and markers are attached. **ACTIONS AT 6-MINUTE WARNING** Issue the 6-minute time warning. Remove personnel restraints or seat belts. Position personnel and equipment. Break chemlights (night operations). Open aircraft doors, if required. **ACTIONS AT 1-MINUTE WARNING** Issue the 1-minute time warning and STAND BY. Position ropers in stick formation. **ACTIONS AT FLARE** Identify the target area. Deploy bundles/equipment (Safeties) and clear ropes. Deploy FRIES ropes (FRM and AFRM). Check ropes to make sure 15 feet of rope are on the surface (FRM).

Table 5-1. FRM helicopter fast-rope operations checklist

ACTIONS FOR DESCENT
~ FRM, AFRM, or safety positions the number 1 man at the rope. (The FRM may exit first or last.) ~ FRM issues the command GO. AFRM echoes GO command, and the ropers exit the aircraft. Safety informs PIC, "ROPERS OUT."

AFRM or safety controls the rate of exit of the ropers. AFRM exits last. Aircrew or safeties observe the last roper. Safety tells the PIC, ALL ROPERS AWAY, after the last roper is on the surface and signals. PIC issues command JETTISON or RECOVER ROPES. Aircrew or safety jettisons or recovers ropes, and issues ROPES CLEAR report to the pilot.

Table 5-1. FRM helicopter fast-rope operations checklist (continued).

5-31. FRIES COMMANDS AND SIGNALS

The following standardized arm-and-hand signals are provided for FRIES operations. Ground-to-aircraft arm-and-hand signals are military standard as found in the Pathfinder field manual 57-38.

a. Time warnings (10-, 6-, and 1-minute) are given by holding up the appropriate number of fingers and verbally sounding off.

b. The signal for GET READY is the same as the standard Airborne signal—arms extended horizontally, fingers extended vertically and joined.

c. The signal for STAND-BY is arms extended downward at a 45-degree angle from the body with hands closed except for index fingers, which are extended, pointing to the floor

d. To execute the signal for DEPLOY ROPES, the FRM or safety forms both hands into fists at mid-thigh level and sweeps arms horizontally outward. At full extension, he opens his hands as if dropping the ropes out of the helicopter and sounds off with DEPLOY ROPES.

e. The signal for GO is arms raised to horizontal, hands closed with index fingers extended pointing to the ropes.

f. The signal for STOP or ABORT is hands closed into fists with arms raised across the forehead .

g. The signal for SIT-DOWN and DON SEAT BELTS is pointing to the floor, followed by moving both fists around waist, as if donning a belt.

h. The signal for OK is hands raised to eye level, index finger and thumb forming a circle with other fingers extended.

Section IX. EMERGENCY ACTIONS

Procedures for correcting an emergency condition that could reasonably be encountered are described in this section. Multiple emergencies, adverse weather, or other unusual conditions may require modification of these procedures. The nature and severity of the emergency dictate the response necessary; therefore, personnel must use sound judgment in determining the correct action to be taken.

5-32. EMERGENCIES BEFORE ROPING STARTS

All personnel sit down, don their seat belts, and take further instructions from the pilot or crew chief.

5-33. EMERGENCIES AFTER ROPING STARTS

Emergency procedures to be followed after roping starts are discussed in this paragraph.

a. **Unsafe Drift or Premature Liftoff.** Anyone who detects the aircraft having drifted off the site must immediately stop training and inform the FRM, AFRM , safety, or PIC.

(1) The FRM, AFRM, safety, or roper stops stick.

(2) The ropers stop descent and lock in.

(3) The FRM or crewman informs the PIC, and guides him in moving the aircraft back on target.

(4) The unit continues operations.

b. **Rope Hung or Snagged (On Ground, Tree, Building).** Anyone who detects the rope being snagged or hung must immediately stop training and inform the FRM, AFRM, safety, crew chief, or PIC.

(1) The safety ensures ropers are off the rope and are clear.

(2) The aircraft descends or lands as needed.

(3) The safety frees the rope and informs the pilot.

(4) The unit resumes the operation.

c. **Lost Communications.** All training and operations must be planned to include the use of the intercom between the PIC or crewmen and the FRM. If the intercom fails, the following arm-and-hand signals can be used until the rope can be cleared and the intercom is restored.

- Wrists crossed diagonally above the forehead with clenched fists to signal stop stick.
- Finger pointed toward the exit and a verbal GO to signal ropers to exit.
- Open palm moved and faced in the required direction to signal aircraft movement.
- Clenched fist to stop aircraft movement.

Note: Training will not be conducted without radio communications between the pilots or aircrew and the FRM and the ground.

Section X. SAFETY

Safety is everyone's responsibility. All personnel involved in FRIES operations are responsible for identifying hazardous situations and preventing injuries to personnel. Anyone who observes an unsafe condition or act is authorized to halt the operation and inform the FRM or the PIC.

5-34. PREFLIGHT AND INFLIGHT

Anyone who detects an inflight emergency will immediately notify the PIC, crew chief, or safety and follow aircrew instructions.

a. Before a FRIES operation, the FRM will coordinate with the PIC both routine and emergency procedures as well as any unique conditions or potential problems. He also discusses the procedures to be used during flight, during roping, and before the aircraft's departure from the roping site. The FRM then discusses the information with all the AFRMs, safeties, and crew chiefs as well. He reviews the time warnings (including arm-and-hand signals) and appropriate actions such as who will deploy ropes, verify rope status, and perform rope-clearing procedures.

b. The mission prebrief covers any aircraft equipment that, if disturbed, could cause an aircraft or fast-rope problem. The problem areas should be made known to all personnel concerned and preventive measures taken to neutralize or eliminate them where possible.

c. The FRM gives the crew chief the final position adjustment over the target, using hand signals.

d. All safety straps and communications cords that are not in use are securely stowed. All safety straps and communication cords in use should be positioned so they do not present a hazard to personnel movement in the aircraft.

e. After the PIC gives the command DEPLOY ROPES, the FRM is responsible for deploying the rope and making sure it is in contact with the surface before ropers exit.

Note: The safety is responsible for observing rope and ropers and for making sure the pilot maintains aircraft position until all ropers are on the surface.

CAUTION

According to AR 95-3, anytime operations occur over water, all personnel will wear an approved personal flotation device as identified in FM 57-220 or USASOC Regulation 350-2.

5-35. DURING ROPING

The following safety procedures apply during roping.

a. During the actual descent, fast-ropers will not use NVGs because they limit depth perception and create a tunnel vision effect.

b. During descent, ropers must watch for obstructions and lower ropers, and must break as necessary.

c. Individual ropers will lock in during emergencies by wrapping one leg around the rope once or twice and standing on the fast rope with the other foot.

d. Ropers must maintain sufficient intervals on the rope to prevent contact with lower ropers. Individual ropers exit when the head of the man in front of them disappears below the ramp.

e. Ropers execute descents at speeds commensurate with their experience and proficiency in FRIES infiltration operations. (For example, less-experienced ropers should descend slowly until they become more proficient.)

f. The number of ropers on each rope at one time must not exceed the safe limit of the rope mount point in use, as follows:

(1) Do not exceed 1,500 pounds per rope on the MH/UH-60.
(2) Do not exceed 1,500 pounds per rope on the MH/UH-47.
(3) Do not exceed 1,500 pounds per rope on the MH/UH-53.
(4) Do not exceed 1,300 pounds per rope on the MH-60G.

5-36. SAFETY BRIEFING

All units must present a safety briefing before conducting FRIES training or operations. The safety briefing must cover the following (as a minimum):

- The planned event (day or night, tower or helicopter, high or low hover, insertion or extraction).
- Actions on the aircraft (or tower).
- Operation (or training) location and any hazards within 500 meters.
- Identification of key personnel, their location, and their duties.
- FRIES equipment and its characteristics.
- FRIES equipment rigging or inspection.
- Roping personnel preparation (shirttails tucked in, chin strap fastened, loose straps taped, gloves and goggles on).
- Arm-and-hand signals.
- Emergencies and emergency procedures.
- Medical procedures (medic with aid bags, backboards, litters, and dedicated vehicle with driver on site).
- Communications requirements (frequencies, call signs, required reports).
- Any other information essential to safety.

5-37. SAFETY REMINDERS

The following safety reminders will be reviewed by all key personnel before conducting operations.

a. Ropers must wear long-sleeve shirts or jackets, full-length pants, laced boots, helmets, protective goggles, and heavy leather gloves.

b. MILES harnesses must not be worn during fast-roping.

c. Ropers will not wear NVGs.

d. Commanders must personally approve the rope site selected for night operations. The FRIES site must be large enough to permit all ropers, upon reaching the surface, to move clear of the rope and for units to consolidate or re-form.

e. All night rope operations must be assessed as medium risk or higher.

f. FRIES descent training is not authorized with a fast rope longer than 90 feet.

g. Ropers must not carry more than 50 pounds of equipment during FRIES descents. The maximum weight of rucksacks is limited to 35 pounds. When rucksacks exceed 35 pounds (or total load exceeds 50 pounds), gear must be lowered (belayed) on a separate system, preferably from the opposite door.

h. Ropers will not carry bulky equipment that could interfere with FRIES operations. Equipment that cannot be carried by the soldier during roping (due to its weight or bulk) can be lowered (belayed) during insertion. During extractions, rucksacks must be attached to the extraction loops of the FRIES rope using snaplinks.

i. Ropers must not make more than 6 combat equipment descents or a total of 10 descents in a 24-hour period.

j. FRIES training must not be conducted if—

- The aircraft is unstable.
- The fast rope is not fully deployed (15 feet of the rope must be on the surface).
- Any obstacle is present that could interfere with the descent of the ropers.
- The rope is frozen or slippery
- Anyone identifies any unsafe condition.

k. Safety personnel must—

- Position themselves so they can observe active FRIES points with out obstructing ropers.
- Conduct final rigging checks of the mount, equipment, and ropes at the 10-minute warning.
- Ensure the aircraft is over the target area and the ropes reach the ground.
- Constantly keep both the PIC and FRM informed—
  - When the aircraft is in position for deployment.
  - When anything is being deployed from the helicopter (ropes, bundles or personnel).
  - When 15 feet of rope are on the surface.
  - When the first roper exits on the fast rope.
  - When all ropers on the surface are clear.
  - When the ropes are recovered or jettisoned.
  - When the aircraft is cleared for flight.
  - When 15 feet of the extraction rope is on the surface.
  - When extraction personnel are attached to the rope and ready for lift off.
  - When extracted personnel are safely lowered to the ground.
  - When extracted personnel are out from under the aircraft.
  - Any other conditions that could impact safe operations.

Note: If the FRM and the safety exit, the crew chief performs rope clearing and or jettisoning duties.

WARNING

Using the word CLEAR tells pilots the aircraft is ready to fly. Before any use of the word CLEAR, FRMs and safeties must ensure that all personnel are well away from the ropes and the ropes have been fully recovered or released (during insertion), or all personnel are securely attached to the ropes (during extraction).

l. FRIES training must not be conducted in densely wooded areas or areas prone to blowing dust, sand, or snow, which can produce a hazard for ropers or the aircraft.

m. FRIES training and operations must not be conducted during severe weather such as rain, sleet, ice, or extreme cold.

n. The possibility of personnel injury during FRIES training and operations is always present. Every soldier involved in FRIES activities must maintain constant vigilance to prevent personnel injury and equipment damage.

CHAPTER 6

EQUIPMENT

This chapter discusses the basic equipment used for rappelling operations (Appendix B). It covers the types of ropes, the procedures for rope selection, and the care for each rope type. It also discusses the two methods of coiling and the different snaplinks used to join the equipment and rope.

Section I. ROPES

Ropes are the most important items of equipment for the rappeller. They provide access down the obstacle or aerial platform, while ensuring individual safety.

6-1. TYPES OF ROPES

Two types of ropes are available to the rappeller—*static* and *dynamic*.

a. **Static Rope.** This rope allows for minimal stretch of the rope. The static rope stretches about 5 to 15 percent at the point of failure and about 2 percent under a working load. The minimum tensile strength for an 11-mm (7/16-inch) static rope for military use is 4,500 pounds. *Kernmantle* static ropes range from 3 millimeters to 11 millimeters in diameter and are constructed similar to the 550-parachute cord. (*Kern* means core and *mantle* means sheath.) The internal core of the rope is constructed of a continuous multifilament-nylon yarn, which is spun into continuous parallel strands. This internal core is then covered with a nylon-braided outer sheath. Due to the internal parallel strand construction of the static rope, there is less spinning and kinking than with a dynamic rope. Static ropes can be used for slings, harnesses, etriers (stirrups), rappelling, rope installations, hauling lines, and where stretch of the rope is not desired.

b. **Dynamic Rope.** This rope allows for stretch within the fibers of the rope, which can be a disadvantage in rappelling, prusik climbing, and other applications. Dynamic ropes are more susceptible to abrasion and wear. They have about a 5 to 10 percent working stretch. The minimum tensile strength for an 11-mm (7/16-inch) rope for military use is 4,500 pounds. Two types of dynamic ropes are nylon-laid and kernmantle:

(1) *Nylon-laid ropes.* Synthetic fibers are the best climbing and rappelling material. Nylon has become the standard material for climbing ropes and has replaced manila, flax, hemp, and sisal. The military mountaineering critical application rope is constructed in a *laid* fashion with continuous multifilament-nylon fiber yarn twisted into strands. Three strands are twisted into a climbing rope of a specific diameter (Figure 6-1). Nylon-laid ropes are easy to inspect for serviceability by twisting the fibers. Laid ropes tend to untwist slightly when under a load, causing kinking and spinning. They are also susceptible to abrasion. The following specifications pertain to the standard military nylon-laid climbing/rappelling rope:

- 36 1/2 meters (120 feet) long.
- 11 millimeters (7/16 inch) wide.
- At least a 4,500-pound tensile strength.
- Right-hand lay.
- One-third stretch factor (at point of failure).

- May lose as much as 15 percent of its strength when wet.
- Weighs 6 pounds when dry.

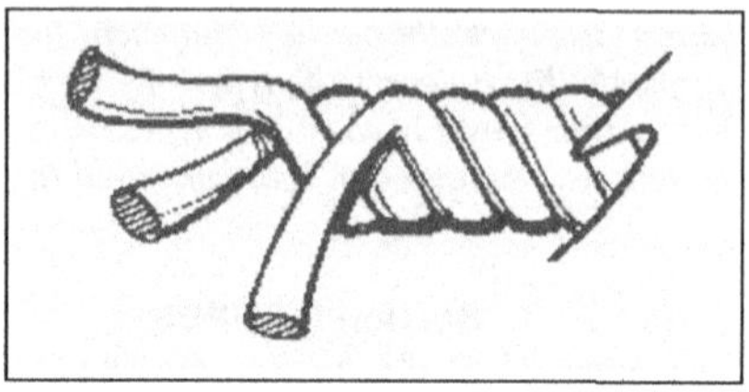

Figure 6-1. Nylon-laid rope.

(2) *Kernmantle ropes.* These ropes are similar to their static counterparts in that they consist of an inner core and an outer sheath. The core is constructed of continuous twisted nylon filaments, which are laid or braided together and enclosed in a tightly braided outer sheath. These ropes are well suited for climbing on rock, snow, or ice where a brief elastic stretch of the rope occurs during a fall. They have a stretch of about 40 percent at the point of failure. The breaking strength is high, and there are no exposed strands for rock crystals to work between to damage the rope (Figure 6-2). There is less sliding friction through the snaplink and over other surfaces, since the outer sheath is smooth. Kernmantle ropes are available in many sizes, lengths, stretch factors, tensile strengths, and fall ratings.

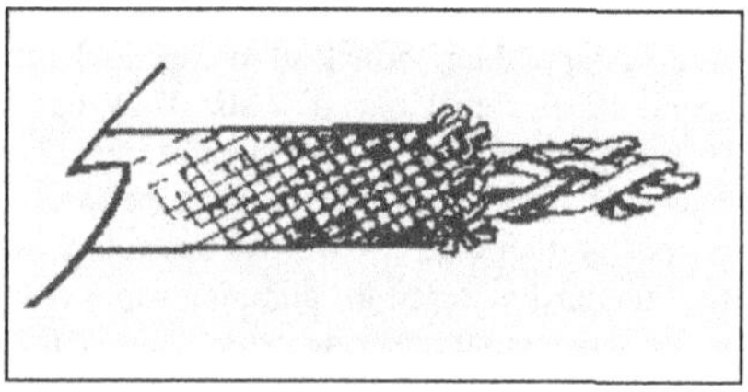

Figure 6-2. Kernmantle (core and sheath) rope.

6-2. ROPE SELECTION

Rope selection for military operations requires ropes of various types, lengths, and diameters; and depends on its intended use (static or dynamic), environment (water-repellant), climate, mission (lightweight ropes), and other factors. Static ropes should have minimal stretch; be smooth, flexible, and strong; resist abrasion and cutting; and have a low-impact force.

a. *Laid rope* is easier to inspect (by twisting the strands apart to observe the interior surfaces), has greater friction in holding knots, and is less expensive than kernmantle ropes. Its disadvantages include: much lower abrasion resistance, extremely high stretch factor for both static and dynamic loads, greater friction on rock, readiness to kink, tendency to spin a free-hanging rappeller, and greater absorption of water and dirt.

b. *Kernmantle* rope is easy to work with, has less friction, does not kink easily, and has less susceptibility to abrasion. The outer sheath protects the internal fibers from dirt, abrasion, and ultraviolet radiation. Its disadvantages are that it is hard to determine damage to the core fibers, and it is more expensive than laid rope. Kernmantle rope is not readily identifiable as dynamic or static and must be marked accordingly.

c. Rope diameter and length vary with intended use. A standard diameter for many uses is 11 millimeters (7/16 inch). For stirrups, utility ropes, and hauling lines, 7 millimeters (5/16 inch), which can also be used in double strands, is adequate. Standards to remember are that the smaller the diameter, the less the tensile strength; but the larger the diameter, the heavier the rope. The standard lengths are 36 1/2 meters (120 feet), 40 meters (135 feet), 45 meters (150 feet), and 50 meters (165 feet). Normally, standard lengths should not be cut. If a nonstandard length is needed, the rope should be precut before the mission and clearly marked.

d. A rope should be selected for a specific use. There are different types of ropes for various uses. A dynamic rope is designed for climbing, while a static rope is designed for rappelling, rescue operations, load hauling, and rope installations.

WARNING

Using a rope for other than its intended use could result in personal injury or damage to equipment.

6-3. CARE OF ROPES

Because the rope is the rappeller's lifeline, it requires a great amount of care and maintenance.

a. Before and after use, the rope should be carefully inspected for excessive wear, cuts, mildew, or rotten spots. If any of these defects are found, the rope is unserviceable. An unserviceable rope is destroyed by cutting it into usable parts. If no sections are usable, the rope is cut into smaller parts and discarded.

b. Ropes should never be spliced together, stepped on, or dragged. This greatly reduces the tensile strength of the rope.

c. The rope must not come in contact with sharp edges. If this is likely, the areas must be padded or taped before rappelling.

d. The rope must be kept as dry as possible. If it becomes wet, it must be dried out to prevent rotting. (Avoid excessive heat/heating during the drying process.)

6-4. COILING

Two methods of coiling are the *mountaineer's coil* and the *farmer's coil* (Figure 6-3).

a. **Mountaineer's Coil.**

(1) To use the mountaineer's coil, grasp the rope about 3 feet from the end with the left hand. The right hand is then placed next to the left hand and runs along the rope until both arms are outstretched. Grasping the rope firmly, bring the hands together to form a loop, which is laid in the left hand. Repeat, forming uniform loops that run in a clockwise direction until the rope is completely coiled. If the rope tends to twist or form figure eights, give it a slight twist with the right hand when each loop is formed. Coil the rope in a clockwise direction to conform with the lay.

(2) To complete the coil, form a bight about 30 centimeters (12 inches) long with the starting end of the rope and lay it along the top of the coil. Uncoil the last loop and use it to wrap around the coil and the bight. Wrap toward the closed end of the bight, binding the final wrap across itself to lock it in place. Make six to eight wraps to secure the coil. Push the end of the rope down through the closed end of the bight. Carry the running end of the coil either in the pack (by forming a figure eight, doubling it, and placing it under the flap) or by placing it over the shoulder and under the opposite arm. If the rope to be coiled is anchored as in coiling a belay or rappel lane, start the coil near the end closest to the anchor so that the kinks work themselves out of the free end.

b. **Farmer's Coil.**

(1) The farmer's coil is easier and faster to coil than the mountaineer's coil. Establish the center of the rope either by locating its center mark or by grasping both ends of the rope and feeding them out until a bight comes up that is the center of the rope. Begin coiling with the doubled rope the same as the mountaineer's coil method. Leave about 4.6 meters (15 feet) of rope uncoiled. Then, squeeze the coils together and make four to six wraps around the middle of the coil, ensuring that the first wrap is held in place by the other wraps.

(2) Form a bight with the two running ends and place it through the bight formed by the top of the coil. Then, run the two running ends over the top of the coil and through the bight that is formed. Dress them in the center of the back of the carrier. Run the two ends over the shoulders to form shoulder straps. Then, bring them under the arms, cross them in the back over the coil, bring them around the body of the carrier, and tie them off with a square knot at the stomach.

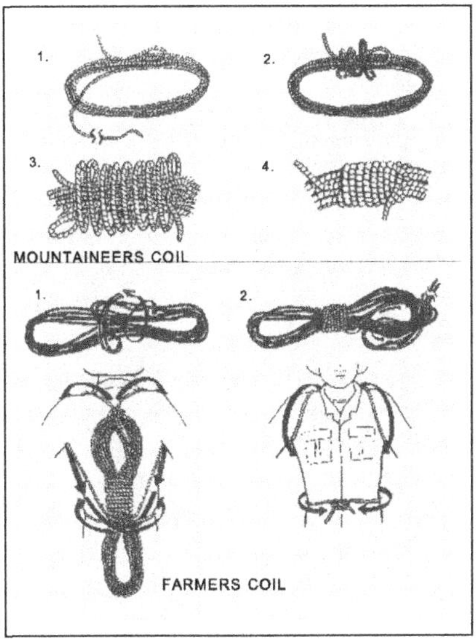

Figure 6-3. Two methods of tying a coil.

Section II. SNAPLINKS

Snaplinks are used to join equipment, rope, and people into a functioning system. When used properly, they are strong, versatile items. Snaplinks are metal (steel, aluminum, or alloys) loops with a hinged, spring-loaded gate on one side. They come in many sizes, shapes, and strengths (with and without locking gates) and in many types of metal. Steel is the strongest snaplink, but it is heavier.

6-5. DESCRIPTION

Snaplinks are available in many sizes and shapes to suit different needs. A locking snaplink is best for rappelling. Nonlocking snaplinks are easy to operate, but should be used only where they cannot be accidentally opened. Hollow snaplinks should be avoided since their use is limited. The following information applies to all snaplinks.

a. The weakest part of a snaplink is the gate. The gate must be closed before applying a load. When the gate is open, snaplinks should have little or no lateral movement of the gate.

b. Locking pins should be checked to ensure that they are not loose, worn, or corroded.

c. The metal should be checked for any cracks, grooves, burrs, flaws, or rust.

d. The spring-loaded gate should automatically close securely from an open to a closed position with no gap between the locking pin and notch.

e. If an engraver is used to mark snaplinks, it should be applied only to the gate, never to the load-bearing side.

f. A snaplink should never be side-loaded (across the gate) since this reduces the overall strength to the point of gate failure.

6-6. TYPES OF SNAPLINKS

The two types of snaplinks used during military operations are the standard snaplink (Figure 6-4) and the locking gate snaplink (Figure 6-5). Both snaplinks are available in a D-shaped variation or an oval variation.

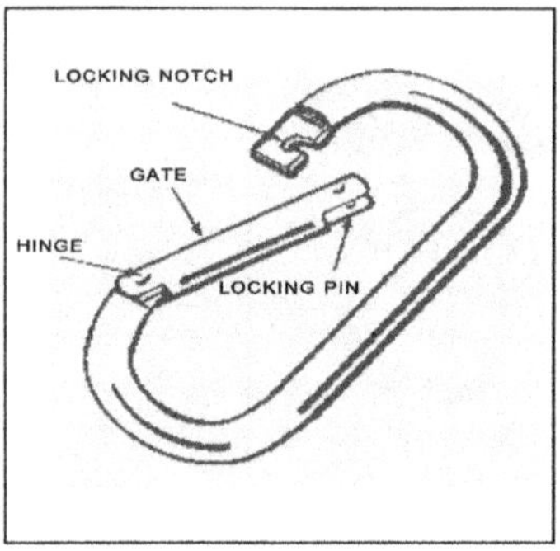

Figure 6-4. Standard snaplink (oval).

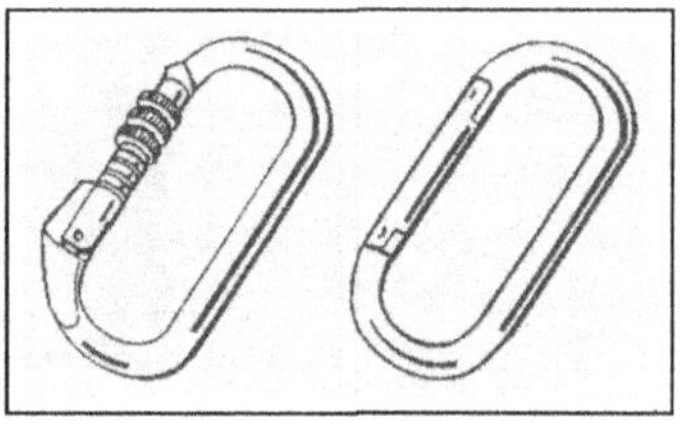

Figure 6-5. D-shaped snaplinks.

a. **D-Shaped Snaplinks.** D-shaped snaplinks are stronger than the oval type because the shape directs the largest part of the load to be applied to the longer, stronger side opposite the gate (Figure 6-5). D-shaped snaplinks are made of steel or aluminum alloys, and are available in many sizes and thicknesses, all of which can be with or without locking gates. Their strength and durability vary, which must be considered before use.

b. **Oval Snaplinks.** Oval snaplinks are versatile and have different applications. They are made of steel or aluminum alloys and are available in many sizes and thicknesses. Both sides of an oval snaplink bear the strain equally under load weight. Many modified ovals are available and most can be with or without the locking gate. All must meet military standards or UIAA specifications.

c. **Locking Snaplinks.** Locking snaplinks have a locking mechanism with a threaded sleeve on the gate. The sleeve screws tightly over the gate opening end or hinge end to hold the gate closed (Figure 6-6). A reverse locking gate is needed to prevent a moving rope from unscrewing the sleeve. The locking sleeve and threads should be kept free of dirt and grit. If the sleeve is forced to close, it may strip the threads. The locking mechanism ages and weakens after repeated use and should be routinely inspected for wear. The characteristics of locking snaplinks are as follows:

- Material: steel.
- Approximate strength: 2,000 pounds with the gate closed.
- Gate: spring-loaded without locking sleeve.
- Weight: 4.3 ounces.

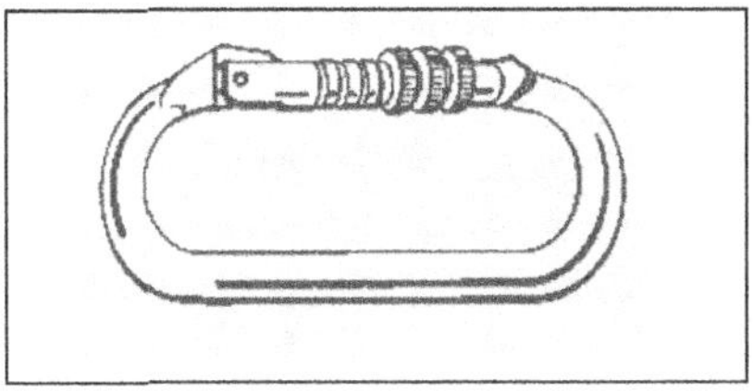

Figure 6-6. Locking snaplink.

6-7. INSPECTION

All rappelling material and individual rappelling equipment must be inspected before conducting a rappelling operation. The rappel master is responsible for ensuring the serviceability of rappelling material and individual rappelling equipment.

a. **Snaplinks.** Snaplinks should be inspected daily before, during, and after use.

(1) The metal should be checked for cracks, grooves, burrs, rust, and flaws. The gate should open and close freely without binding. There should be no lateral movement when the gate is open. The gate spring action should snap shut when released. The locking notch should have a slant or slot so that the gate remains shut under the impact of a rappeller's fall. The gate pins should not work their way out of their holes and should not be shorter than their holes. If there is a locking mechanism, it should be inspected to ensure that threads are not stripped and that the sleeve tightly locks the gate.

(2) If burrs, grooves, or rough areas are identified, the snaplink should not be used. Rust should be removed with steel wool and the spot rubbed with oil or solvent. The spring should be lubricated as needed. The snaplink must be boiled in water for 20 to 30 seconds to remove the cleaning agents since solvents and oils cause dirt to cling to the snaplink and to rub off on the ropes. It is better to use a dry graphite-based lubricant on snaplinks since such lubricants do not attract dirt.

b. **Ropes.** The ropes are inspected for serviceability. A rappel rope is unserviceable if it is saturated with petroleum products, is mildewed, shows excessive wear, has soft spots (internal damage), or if one strand is cut more than one-half of its diameter. Ropes should be inspected daily—before, during, and after use. An applicable annotation should be made in a corresponding rope log (DA Form 5752-R) (Figure 6-7).

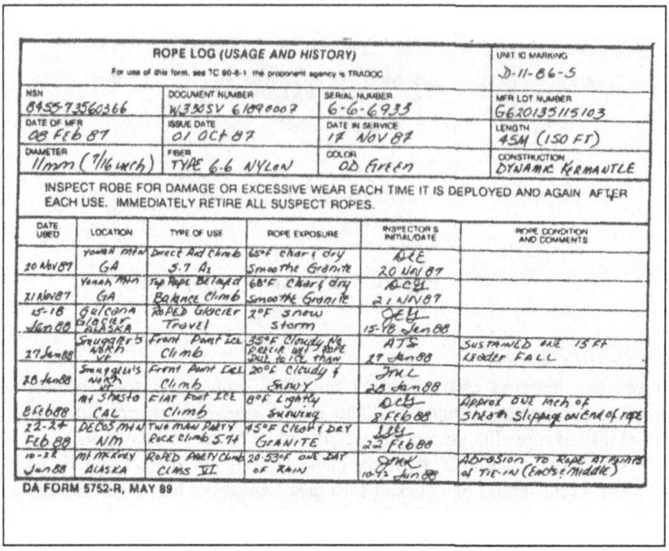

ROPE LOG (USAGE AND HISTORY)

For use of this form, see TC 90-6-1; the proponent agency is TRADOC

NSN 8455-73560366	DOCUMENT NUMBER W330SV 61890007	SERIAL NUMBER 6-6-6933	UNIT ID MARKING D-11-86-5 MFR LOT NUMBER G620135115103
DATE OF MFR 08 FEB 87	ISSUE DATE 01 OCT 87	DATE IN SERVICE 17 NOV 87	LENGTH 45M (150 FT)
DIAMETER 11mm (7/16 inch)	FIBER TYPE 6.6 NYLON	COLOR OD Green	CONSTRUCTION DYNAMIC KERMANTLE

INSPECT ROBE FOR DAMAGE OR EXCESSIVE WEAR EACH TIME IT IS DEPLOYED AND AGAIN AFTER EACH USE. IMMEDIATELY RETIRE ALL SUSPECT ROPES.

DATE USED	LOCATION	TYPE OF USE	ROPE EXPOSURE	INSPECTOR'S INITIAL/DATE	ROPE CONDITION AND COMMENTS
20 Nov 87	Yonah Mtn GA	Direct Aid Climb 5.7 A1	65°F clear & dry Smoothe Granite	DLE 20 Nov 87	
21 Nov 87	Yonah Mtn GA	Top Rope Belayed Balance Climb	68°F Clear & dry Smoothe Granite	DCY 21 Nov 87	
15-18 Jan 88	Gulcana Glacier ALASKA	ROPED Glacier Travel	2°F snow storm	JEY 15-18 Jan 88	
27 Jan 88	Smugglers North VT	Front Point Ice Climb	35°F Cloudy No Rescue wet rope Due to ice thaw	ATS 27 Jan 88	SUSTAINED ONE 15 FT Leader FALL
28 Jan 88	Smugglers North VT	Front Point Ice Climb	20°F Cloudy & Snowy	JML 28 Jan 88	
8 Feb 88	Mt Shasta CAL	Flat Foot Ice Climb	8°F Lightly Snowing	DCY 8 Feb 88	Approx one inch of sheath slippage on end of rope
22-24 Feb 88	DECOS Mtn NM	Two man Party Rock climb 5.7+	45°F Clear & DRY GRANITE	JLY 22 Feb 88	
10-32 Jun 88	Mt McKinley ALASKA	ROPED PARTY Climb CLASS VI	20-53°F one DAY of RAIN	JMK 10-32 Jun 88	Abrasion to Rope at points of tie-in (Ends & middle)

DA FORM 5752-R, MAY 89

Figure 6-7. Example of completed DA Form 5752-R.

c. **Gloves.** The gloves are inspected for serviceability. They are unserviceable if holes are found in the friction-bearing surfaces or along the seams.

d. **Helmet.** The helmet is inspected for correct assembly and serviceability. It is unserviceable if it cannot be properly assembled or if any straps are cut or torn one-quarter of its width.

(1) Ensure that the chin strap attaching tabs are secured to the helmet by the locking nuts.

(2) Ensure that the pull-the-dot tab has four plies of nylon and that the chin strap is routed from the inside of the buckle around the horizontal bar and back toward the inside of the buckle.

Section III. ALTERNATE METHODS OF DESCENT

Snaplinks or carabiners that are used for rappelling are placed through the sewn loops of the seat harness or the sewn loops of the seat-chest combination.

6-8. FIGURE-EIGHT DESCENDER

The rappeller attaches a locking carabiner to the harness (Figure 6-8). He routes the rappel rope up through the large hole and places the bight over the collar. He inserts the figure eight with the rappel rope attached into the locking carabiner, and locks it down. The rappeller brakes to the rear and descends as in a seat hip rappel. If heavy loads are to be descended, he uses a double wrap around the collar.

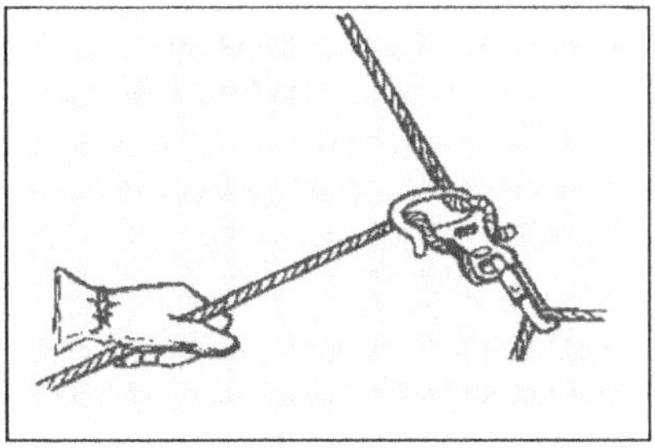

Figure 6-8. Figure-eight descender.

6-9. MUNTER HITCH

The rappeller attaches a large radius snaplink to the harness or rappel seat. He ties a Munter hitch and clip into the snaplink (Figure 6-9, page 6-9). He brakes to the front and descends with an L-shape body position.

Notes: 1. The guide hand should remain on the standing end of the rope to keep the rappeller's head and face away from the hardware.

2. The Munter hitch creates significant rope-to-rope friction and may cause premature wear on nylon ropes.

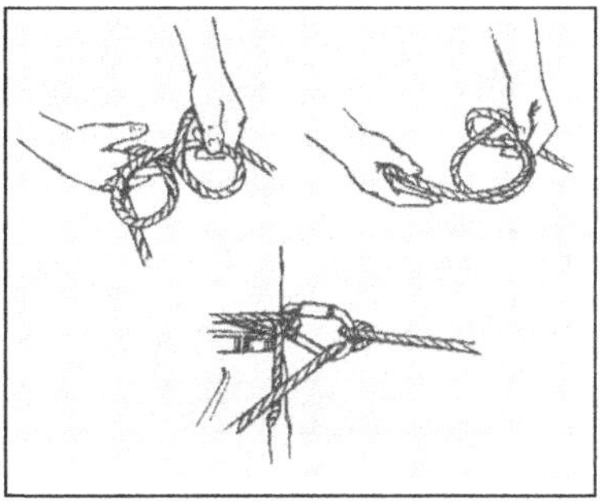

Figure 6-9. Munter hitch.

CHAPTER 7

KNOTS

A knot is a fastening made by intertwining or tying pieces of string, cord, rope, or webbing. All knots used by a rappeller are divided into four classes: Class I—joining knots, Class II—anchor knots, Class III—middle rope knots, and Class IV—special knots. These classes of knots are intended only as a general guide since some of the knots discussed herein may be appropriate in more than one class. The skill of knot tying may be lost if it is not used and practiced. With experience and practice, knot tying becomes instinctive and helps the soldier in many situations.

7-1. MILITARY MOUNTAINEERING TERMINOLOGY

The terms explained below are the most commonly used in military mountaineering (Figure 7-1).

a. A *bight of rope* is a simple bend of rope in which the rope does not cross itself.

b. A *loop* is a bend of rope in which the rope does cross itself.

c. A *half hitch* is a loop that runs around an object and locks itself.

d. The *running end* (working end) of the rope is the free end of the rope that can be used.

e. The *standing part* of the rope is the part that is static (anchored, coiled); the remaining part of the rope not being used (also called static end).

f. The *lay* of the rope is the same as the twist of the rope.

g. A *round turn* is a single complete wrap of the rope around an object providing 360-degree contact. The running end leaves the completed circle in the same direction as the standing part. In a round turn, the rope is wrapped around an object 1 1/2 times.

h. The *pigtail* is the short length of rope remaining at the end after tying a knot or coiling a rope.

i. *Back feeding* (or stacking) is taking off one wrap at a time from a coil, and letting it fall naturally to the ground.

j. *Dress down the knot* is tightening down a knot to its functioning form.

k. An *overhand* is a loop with the running end pulled through the loop.

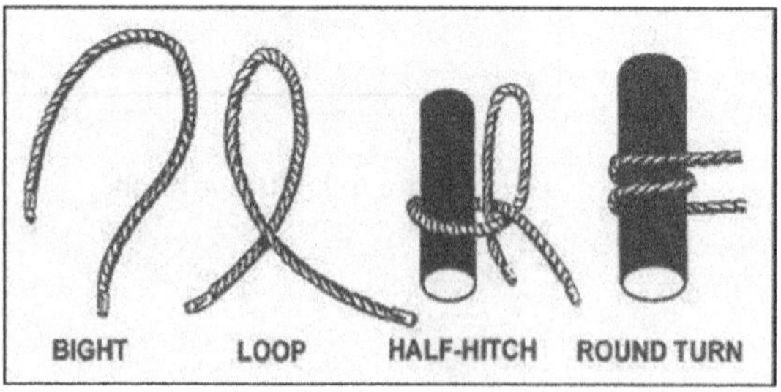

Figure 7-1. Examples of rope terminology.

Section I. JOINING KNOTS

7-2. SQUARE KNOT

The square knot is used to tie the ends of two ropes of equal diameter (Figure 7-2).

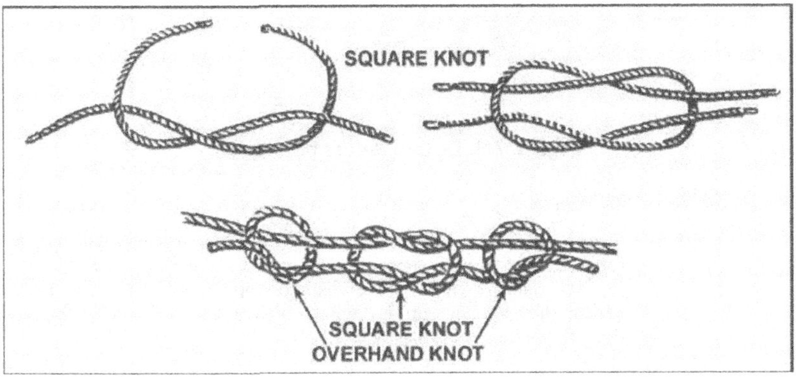

Figure 7-2. Square knot.

a. **Tying the Knot.**

STEP 1. Holding one working end in each hand, place the working end in the right hand over the one in the left hand.

STEP 2. Pull it under and back over the top of the rope in the left hand.

STEP 3. Place the working end in the left hand over the one in the right hand and repeat STEP 2.

STEP 4. Dress the knot down and secure it with an overhand knot on each side of the square knot.

b. **Checkpoints.**

(1) There are two interlocking bights.

(2) The standing parts are on the same side and properly secured with overhand knots.

7-3. DOUBLE SHEET BEND KNOT

The double sheet bend knot is used to tie the ends of two or more ropes of equal or unequal diameter (Figure 7-3). When a single rope is tied to a multiple of ropes, the bight is formed with the multiple of ropes.

Figure 7-3. Double sheet bend knot.

a. **Tying the Knot.**

STEP 1. Form a 30-centimeter (12-inch) bight in the left hand with the rope or ropes. Ensure that the short end of the bight is facing to the inside.
STEP 2. Place the index finger of the left hand on top of the bight. Bring the rope in the right hand up through the bight and over the index finger of the left hand, so the working end is away from the body.
STEP 3. Reach through the loop and grasp the working end of the rope and bring it back towards the body. Place it between the index finger and the bight, forming a round turn. Repeat this one more time, wrapping toward the tip of the index finger.
STEP 4. Remove the finger and dress the knot down.

b. **Checkpoints.**

(1) The two wraps around the bight are held in place by a locking bar.

(2) The two standing parts of the ropes exit parallel and from opposite ends of the knot. The two working ends form an L and exit the knot at a 90-degree angle and on opposite sides of the knot.

7-4. FISHERMAN'S KNOT

The fisherman's knot is used to tie two ropes of similar or dissimilar materials (Figure 7-4).

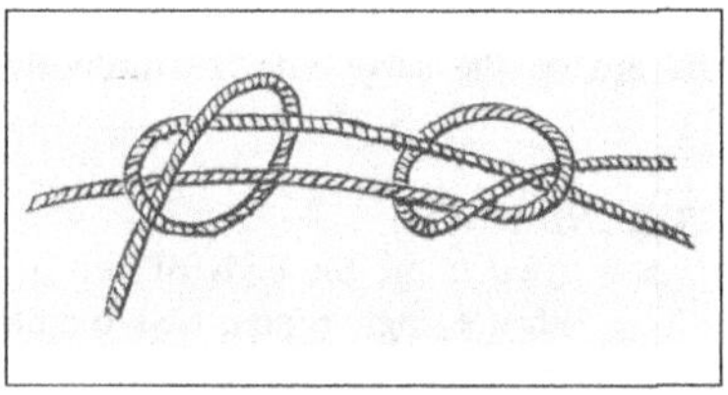

Figure 7-4. Fisherman's knot.

a. **Tying the Knot.**

STEP 1. Tie an overhand knot in one end of the rope.
STEP 2. Pass the working end of the other rope through the first overhand knot. Tie an overhand knot around the standing part of the first rope with the working end of the second rope.
STEP 3. Tightly dress down each overhand knot and tightly draw the knots together.

b. **Checkpoints.**

(1) The two separate overhand knots are tied tightly around the long standing part of the opposing rope.

(2) The two overhand knots are drawn snug.

7-5. DOUBLE FISHERMAN'S KNOT

The double fisherman's knot (also called double English or grapevine) is used to tie two ropes of similar or dissimilar materials (Figure 7-5).

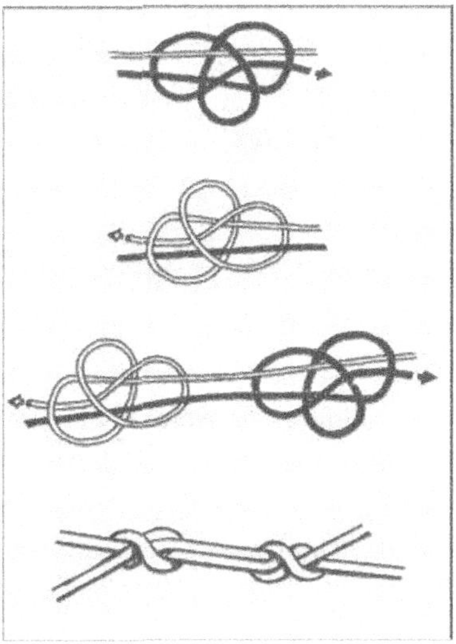

Figure 7-5. Double fisherman's knot.

a. **Tying the Knot.**

STEP 1. With the working end of one rope, tie two wraps around the standing part of another rope.
STEP 2. Insert the working end (STEP 1) back through the two wraps and draw it tight.

STEP 3. With the working end of the other rope, which contains the standing part (STEPS 1 and 2), tie two wraps around the standing part of the other rope (the working end in STEP 1). Insert the working end back through the two wraps and draw tight.

STEP 4. Pull on the opposing ends to bring the two knots together.

b. **Checkpoints.**

(1) Two wraps with the working end running through and drawn tight.

(2) Both wraps are drawn together with all the slack out of the knot.

7-6. WATER KNOT

The water knot is used to attach two webbing ends (Figure 7-6). It is also called a ring bend, overhand retrace, or tape knot. It is used in runners and harnesses.

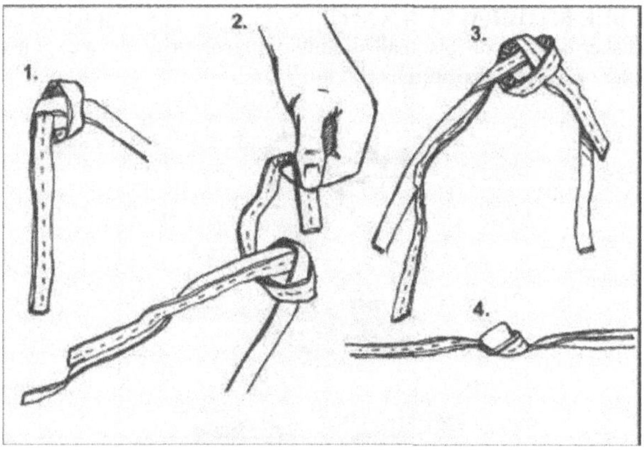

Figure 7-6. Water knot

a. **Tying the Knot.**

STEP 1. Tie an overhand knot in one of the ends.

STEP 2. Feed the other end back through the knot, following the path of the first rope in reverse.

b. **Checkpoints.**

(1) There are two overhand knots, one retracing the other.

(2) There is no slack in the knot, with the working ends coming out of the knot in opposite directions (at least 5 centimeters [2 inches] long).

Section II. ANCHOR KNOTS

7-7. BOWLINE KNOT

The bowline knot is used to tie a single fixed loop in the end of a rope (Figure 7-7). It does not slip under strain and is easily untied. This knot is always used when there is alternating tension. It can also be used to tie the end of a rope to an anchor.

Figure 7-7. Bowline knot.

a. **Tying the Knot.**

STEP 1. Bring the working end of the rope around the anchor, from right to left (as the rappeller faces the anchor).
STEP 2. Form an overhand loop in the standing part of the rope (on the rappeller's right) toward the anchor.
STEP 3. Reach through the loop and pull up a bight.
STEP 4. Place the working end of the rope (on the rappeller's left) through the bight, and bring it back onto itself. Dress the knot down.
STEP 5. Form an overhand knot with the tail from the bight.

b. **Checkpoints.**

(1) The loop is locked into place by a bight.

(2) The short portion of the bight is on the inside and on the loop around the anchor (or inside the fixed loop).

7-8. ROUND TURN AND TWO HALF HITCHES KNOT The round turn and two half hitches knot is used to tie the end of a rope to an anchor and must have constant tension (Figure 7-8).

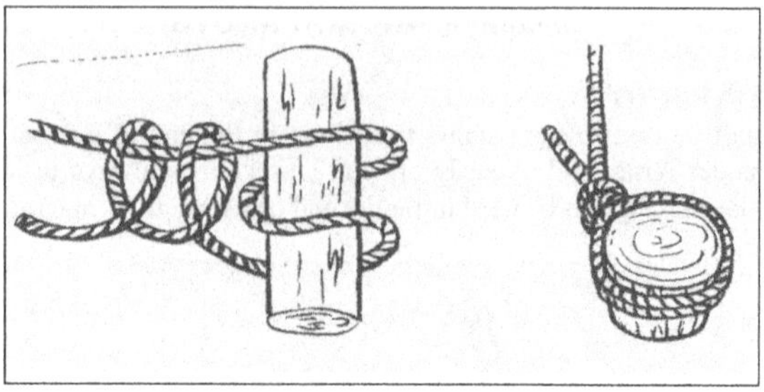

Figure 7-8. Round turn and two half hitches knot.

a. **Tying the Knot.**

STEP 1. Route the rope around the anchor from right to left and wrap down (must have two wraps in the rear of the anchor and one in the front). Run the loop around the object to provide 360-degree contact, distributing the load over the anchor.

STEP 2. Bring the working end of the rope left to right and over the standing part, forming a half hitch (first half hitch).

STEP 3. Repeat STEP 2 (last half hitch has a 15-centimeter [6-inch] pigtail).

STEP 4. Dress the knot down.

b. **Checkpoints.**

(1) A complete round turn should exist around the anchor with no crosses.

(2) Two half hitches should be held in place by a locking bar with no less than a 15-centimeter (6-inch) tail remaining.

(3) When the two half hitches are twisted to the left or the right, they will appear to for a figure eight.

(4) Looking at the top of the half hitches, the rope should appear to be one continuous rope.

7-9. FIGURE-EIGHT RETRACE KNOT

The figure-eight retrace knot is also called a rerouted figure eight. Although it produces the same result as a figure-eight loop, by tying the knot in a retrace it can be used to fasten the rope to trees or to places where the loop cannot be used (Figure 7-9).

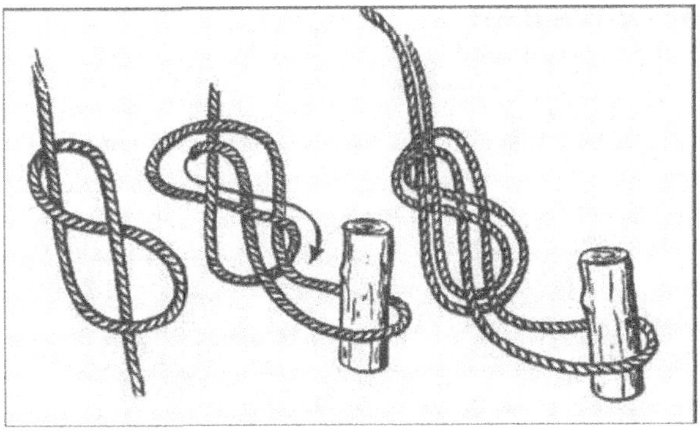

Figure 7-9. Figure-eight retrace.

a. **Tying the Knot.**

STEP 1. Use a length of rope long enough to go around the anchor, leaving enough rope to work with.

STEP 2. Tie a figure-eight knot in the standing part of the rope, leaving enough rope to go around the anchor. To tie a figure-eight knot:

- Form a loop in the rope.
- Wrap the working end around the standing part.
- Route the working end through the loop.
- The finished knot is dressed loosely.

STEP 3. Place the working end around the anchor point.

STEP 4. With the working end, insert the rope back through the loop of the knot in reverse.

STEP 5. Keep the original figure eight as the outside rope and retrace the knot around the wrap and back to the long sttanding part.

STEP 6. Safety off with an overhand knot.

b. **Checkpoints.**

(1) At least 15 centimeters (6 inches) of rope extends past the figure-eight knot.

(2) The original figure eight, tied with the standing end, remains as the outermost rope throughout the knot.

(3) The knot is dressed tightly.

7-10. CLOVE HITCH KNOT

The clove hitch knot can be used in the middle of the rope as well as at the end of the rope (Figure 7-10). The knot must have constant tension on it once tied to prevent slipping. It can be used as either an anchor or middle knot, depending on how it is tied. (Do not use a clove hitch knot on metal except on a picket-hold-fast.)

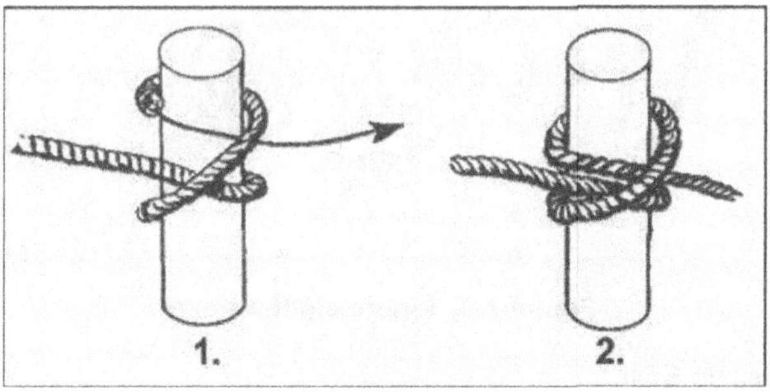

Figure 7-10. Clove hitch knot.

a. **Tying the Knot.**

(1) If there is access over the top of the anchor, the knot is tied as follows:

STEP 1. Hold rope in both hands, palms down with hands together. Slide the left hand to the left from 20 to 25 centimeters (8 to 10 inches).

STEP 2. Form a loop away from and back toward the right.

STEP 3. Slide the right hand from 20 to 25 centimeters (8 to 10 inches) to the right. Form a loop inward and back to the left hand.

STEP 4. Place the left loop on top of the right loop. Place both loops over the anchor and pull both ends of the rope in opposite directions to complete the knot.

(2) If the clove hitch knot must go around the anchor, tie the knot as follows (for instructional purposes, assume that the anchor is horizontal):

STEP 1. Place 76 centimeters (30 inches) of rope over the top of the anchor. Hold the standing end in the left hand. With the right hand, reach under the horizontal anchor, grasp the working end, and bring it inward.

STEP 2. Place the working end of the rope over the standing end (to form a loop). Hold the loop in the left hand. Place the working end over the anchor from 20 to 25 centimeters (8 to 10 inches) to the left of the loop.

STEP 3. With the right hand, reach down to the left-hand side of the loop under the anchor. Grasp the working end of the rope and bring it up and outward.

STEP 4. Dress down the knot.

b. **Checkpoints.**

(1) The knot has two round turns around the anchor with locking bar.
(2) The locking bar is facing 90 degrees from the direction of pull.
(3) The ends exit 180 degrees from each other.
(4) The knot has more than a 15-centimeter (6-inch) pigtail remaining.

SECTION III. MIDDLE ROPE KNOTS

7-11. WIREMAN'S KNOT

The wireman's knot forms a single, fixed loop in the middle of the rope (Figure 7-11).

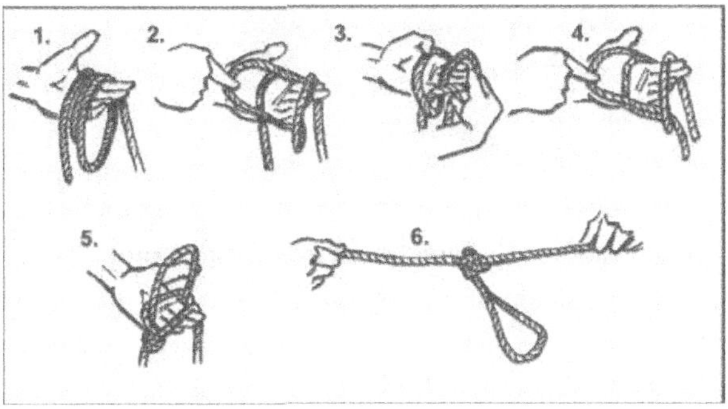

Figure 7-11. Wireman's knot.

a. **Tying the Knot.**

STEP 1. When tying this knot, face the anchor that the tie-off system will be tied to. Take up the slack from the anchor, and wrap two turns around the left hand (palm up) from left to right.
STEP 2. Take up a loop of 30 centimeters (12 inches) in the second round turn to create the fixed loop of the knot.
STEP 3. Name the wraps from the palm to the fingertips: heel, palm, and fingertip.
STEP 4. Grasp the palm wrap with the right thumb and forefinger, and place it over the heel wrap.
STEP 5. Grasp the heel wrap and place it over the fingertip wrap.
STEP 6. Grasp the fingertip wrap and place it over the palm wrap.
STEP 7. Grasp the palm wrap and pull up to form a fixed loop.
STEP 8. Dress the knot down by pulling on the fixed loop and the two working ends.
STEP 9. Pull the working ends apart to finish the knot.

b. **Checkpoints.**

(1) The completed knot should have four separate bights locking down on themselves with the fixed loop exiting from the top of the knot and laying toward the near side anchor point.

(2) Both ends should exit opposite each other without any bends.

7-12. DIRECTIONAL FIGURE-EIGHT KNOT

The directional figure-eight knot forms a single, fixed loop in the middle of the rope that lays back along the standing part of the rope (Figure 7-12).

Figure 7-12. Directional figure-eight knot.

a. **Tying the Knot.**

STEP 1. Face the far side anchor so that when the knot is tied, it lays inward.
STEP 2. Lay the rope from the far side anchor over the left palm. Make one wrap around the palm.
STEP 3. With the wrap thus formed, tie a figure-eight knot around the standing part that leads to the far side anchor.
STEP 4. Dress the knot down with the tail and the bight together.

b. **Checkpoints.**

(1) The loop should be large enough to accept a snaplink but no larger than a helmet-size loop.

(2) The tail and bight must be together.

(3) The figure eight is tied tightly.

(4) The bight in the knot faces back toward the near side.

7-13. BOWLINE-ON-A-BIGHT KNOT

The bowline-on-a-bight knot is used to form two fixed loops in the middle of a rope (Figure 7-13).

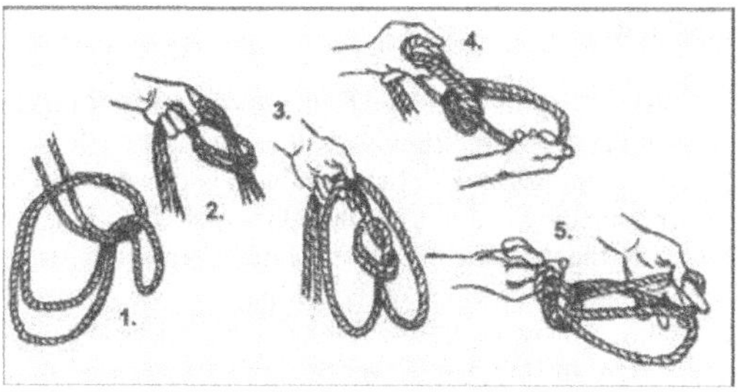

Figure 7-13. Bowline-on-a-bight knot.

a. **Tying the Knot.**

STEP 1. Form a bight in the rope about twice as long as the finished loops will be.
STEP 2. Tie an overhand knot on a bight.
STEP 3. Hold the overhand knot in the left hand so that the bight is running down and outward.
STEP 4. Grasp the bight with the right hand, fold it back over the overhand knot so that the overhand knot goes through the bight.
STEP 5. From the end (apex) of the bight, follow the bight back to where it forms the cross in the overhand knot. Grasp the two ropes that run down and outward and pull up, forming two loops.
STEP 6. Pull the two ropes out of the overhand knot and dress the knot down.
STEP 7. A final dress is required: grasp the ends of the two fixed loops and pull, spreading them apart to ensure the loops do not slip.

b. **Checkpoints.**

(1) There are two fixed loops that will not slip.
(2) There are no twists in the knot.
(3) A double loop is held in place by a bight.

7-14. FIGURE-EIGHT-ON-A-BIGHT KNOT

The figure-eight-on-a-bight knot is used to form two fixed loops in the middle of a rope . (Figure 7-14).

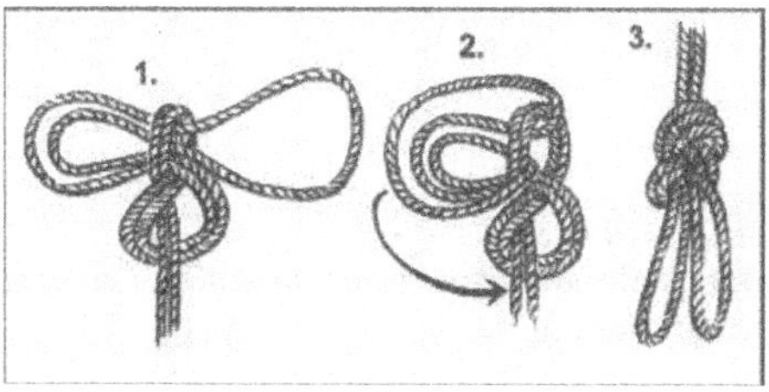

Figure 7-14. Figure-eight-on-a-bight knot.

a. **Tying the Knot.**

STEP 1. Using the doubled rope, form a 46-centimeter (18-inch) bight in the left hand with the running end facing to the left.
STEP 2. Grasp the bight with the right hand and make a 360-degree turn around the standing end in a counterclockwise direction.
STEP 3. With the working end, form another bight and place that bight through the loop just formed in the left hand.
STEP 4. Hold the bight with the left hand, and place the original bight (moving toward the left hand) over the knot.
STEP 5. Dress the knot down.

b. **Checkpoints.**
(1) There is a double figure eight with two loops.
(2) The knot will not slip.

7-15. OVERHAND LOOP KNOT

The overhand loop knot forms a single loop in the middle of a rope. It should not be used in a transport tightening system (Figure 7-15).

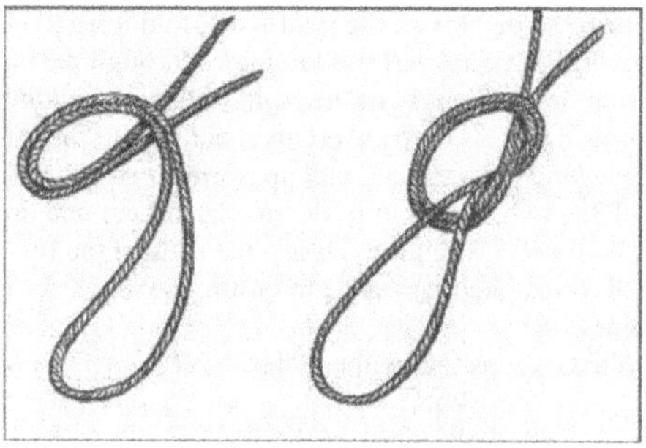

Figure 7-15. Overhand loop knot

a. **Tying the Knot.**
STEP 1. Form a bight in the rope about as long as the loop needed.
STEP 2. With the bight, form an overhand knot using the bight as the working end. Dress the knot tightly.

b. **Checkpoints.**
(1) The loop is the desired size.
(2) The ropes in the overhand knot are parallel and do not cross over each other.
(3) The knot is tightly dressed.

7-16. FIGURE-EIGHT LOOP KNOT

The figure-eight loop knot is a variation of the overhand loop and is easy to untie after it has held weight (Figure 7-16).

Figure 7-16. Figure eight loop knot.

a. **Tying the Knot.**

STEP 1. Form a bight in the rope about as large as the diameter of the desired loop.

STEP 2. With the bight as the working end, form a loop in rope (standing part).

STEP 3. Wrap the working end around the standing part 360 degrees and feed the working end through the loop. Dress the knot tightly.

b. **Checkpoints.**

(1) The loop is the desired size.

(2) The ropes in the loop are parallel and do not cross over each other.

(3) The knot is tightly dressed.

SECTION IV. SPECIAL KNOTS

7-17. SINGLE BUTTERFLY KNOT

The single butterfly knot is used to form a single, fixed loop in the middle of the rope without using the ends (Figure 7-17). The butterfly can be used for the middle man in a rope party as well as in a transport tightening system. The knot can be hard to untie when heavy weight has been placed on it for extended periods.

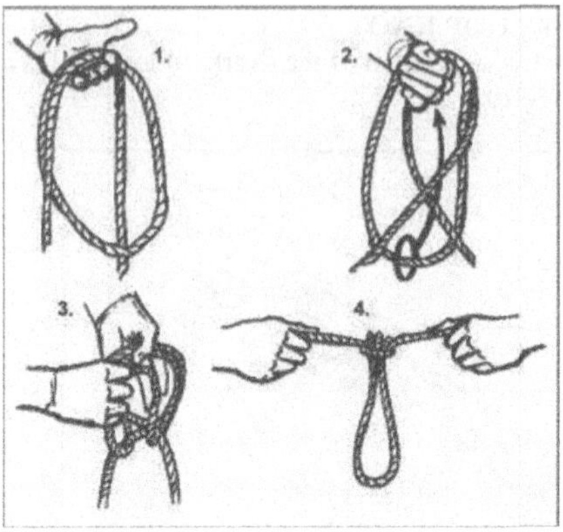

Figure 7-17. Single butterfly knot.

a. **Tying the Knot.**

STEP 1. Form a loop over the palm of the left hand.
STEP 2. With the right hand, cross the two long-standing ends; reach under and grasp the bottom of the loop formed in STEP 1.
STEP 3. Route the bottom of the loop over the crossed standing ends and back through the top of the loop.
STEP 4. Hold the newly formed loop and dress the knot down, removing all of the slack out of the knot.

b. **Checkpoints.**

(1) The wings of the knot are dressed down tightly and as close together as possible.
(2) The ropes between the wings are parallel with no crossovers.
(3) The loop is no larger than a helmet but large enough to accept a snaplink.
(4) All ropes in the knot are tightly dressed.

7-18. PRUSIK KNOT

The prusik knot is used to put a moveable rope on a fixed rope such as a prusik ascent or a tightening system. This knot can be tied as a middle, finger, or end knot.

a. **Tying the Knot.**

(1) *First method.* The prusik knot is tied at the middle or end of a rope; there are many ways to do this (Figure 7-18). The middle-of-the-rope prusik knot can be tied with a short rope to a long rope as follows:

STEP 1. Double the short rope, forming a bight, with the working ends even. Lay it over the long rope so that the closed end of the bight is 30 centimeters (12 inches) below the long rope and the remaining part of the rope (working ends) is the closest to the rappeller; spread the working end apart.

STEP 2. Reach down through the 30-centimeter (12-inch) bight. Pull up both of the working ends and lay them over the long rope. Repeat this process making sure that the working ends pass in the middle of the first two wraps. Now there are four wraps and a locking bar working across them on the long rope.

STEP 3. Dress the wraps and locking bar down to ensure they are tight and not twisted. Tie an overhand knot with both ropes to prevent the knot from slipping during periods of variable tension.

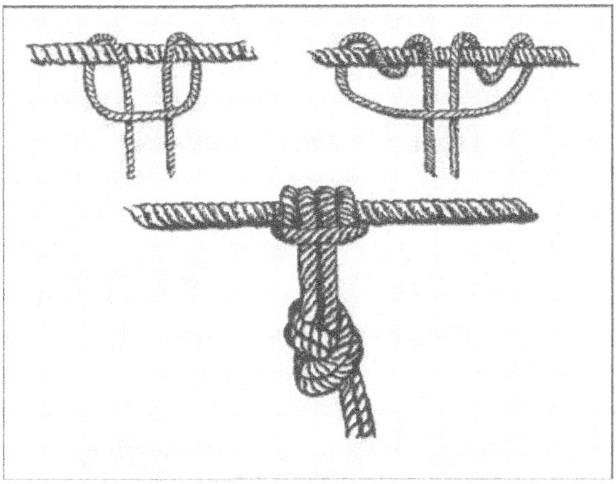

Figure 7-18. Middle-of-the-rope prusik knot.

(2) *Second method.* Another way of tying the prusik knot is the finger prusik (Figure 7-19).

STEP 1. Lay the rope across the thumb and index finger of the left hand.

STEP 2. Wrap the rope around the index finger twice (wrapping away from the rappeller). Wrap the rope around the thumb twice (wrapping towards the rappeller).

STEP 3. Place the thumb and index finger tip to tip. Push the wraps together so that the prusik is on the left thumb. Four wraps and a locking bar should be across them.

STEP 4. Pass the running end of the rope through the wraps of the prusik knot.

STEP 5. Dress the wraps and locking bar down to ensure they are tight and not twisted.

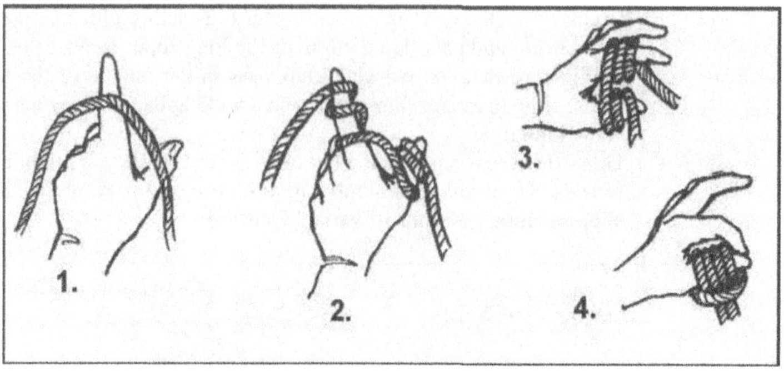

Figure 7-19. Finger prusik knot.

(3) *Third method.* Another way of tying a prusik knot is the end-of-the-rope knot (Figure 7-20).

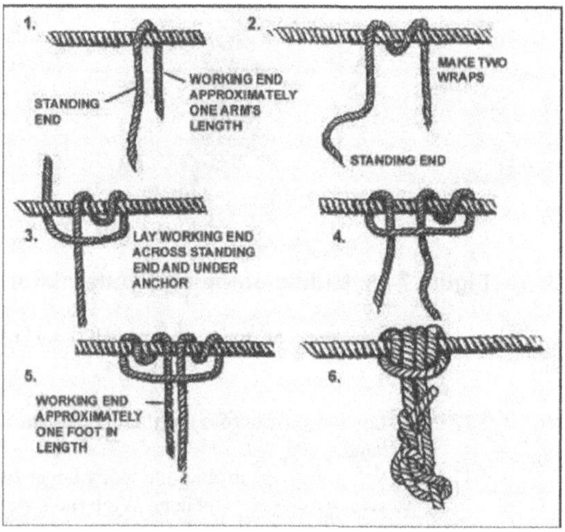

Figure 7-20. End-of-the-rope prusik knot.

STEP 1. Using an arm's length of rope, place it over the long rope.
STEP 2. Form a complete round turn in the rope.
STEP 3. Cross over the standing part of the short rope with the working end of the short rope.
STEP 4. Lay the working end under the long rope.
STEP 5. Form a complete round turn in the rope, working back toward the middle of the knot.

STEP 6. Four wraps and a locking bar should run across them on the long rope. Dress the wraps and locking bar down. Ensure they are tight, parallel, and not twisted.

STEP 7. Finish the knot with a bowline to ensure that the prusik knot will not slip out during periods of varying tension.

b. **Checkpoints.**

(1) Four wraps with a locking bar.

(2) The locking bar faces the rappeller.

(3) The knot is tight and dressed down with no ropes twisted or crossed.

(4) Other than a finger prusik, the knot should contain an overhand or bowline to prevent slipping.

7-19. BACHMAN KNOT

The bachman knot provides a means of using a makeshift mechanized ascender (Figure 7-21).

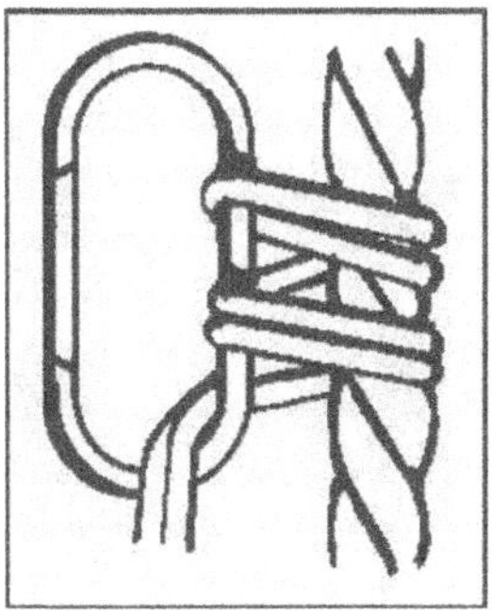

Figure 7-21. Bachman knot.

a. **Tying the Knot.**

STEP 1. Find the middle of a utility rope and insert it into a snaplink.

STEP 2. Place the snaplink and utility rope next to a long climbing rope.

STEP 3. With the two ropes parallel fromthe snaplink, make two or more wraps wraps around the climbing rope and through the inside portion of the snaplink.

Note: The rope can be tied into an etrier (stirrup) and used as a prusik-friction principal ascender.

b. **Checkpoints.**

(1) The bight of the climbing rope is at the top of the snaplink.

(2) The two ropes run parallel without twisting or crossing.

(3) Two or more wraps are made around the long climbing rope and through the inside portion of the snaplink.

7-20. BOWLINE-ON-A-COIL KNOT

The bowline-on-a-coil knot is an expedient tie-in used by rappellers in two- and three-man party climbs when a climbing harness is not available (Figure 7-22).

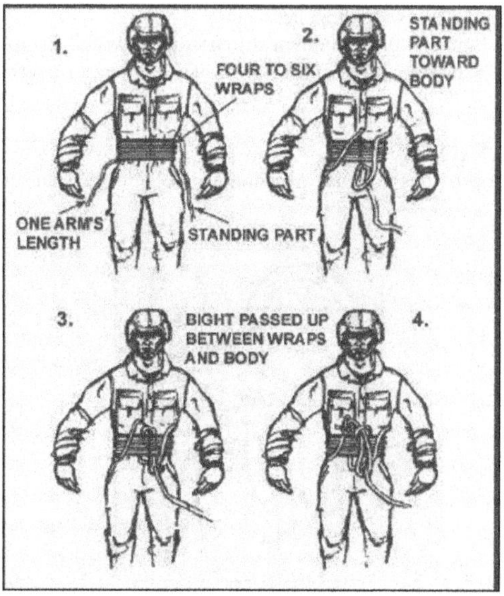

Figure 7-22. Bowline-on-a-coil knot.

Note: A safer tie in, when possible, is a swiss seat and figure eight around the shoulder, to avoid damaging the ribs if the climber falls.

a. **Tying the Knot.**

STEP 1. To tie a bowline-on-a-coil knot, start the same as tying a bowline around the waist. (The waist of the rappeller is the anchor point. The rappeller has about 40 centimeters [16 inches] of rope in the right hand. The standing end is coming from the remainder of the rope in his left hand.)

STEP 2. Draw slack from the standing end of the rope in the left hand, wrapping enough coils around the body to complete a minimum of four (but no more than six) coils.

STEP 3. With the rope in the left hand, make a loop inward with the standing end.

STEP 4. Slide the loop up between the horizontal wraps and the body, ensuring that the loop does not come uncrossed.

STEP 5. Bring a bight up through the loop. Bring the working end of the rope in the right hand through the bight and back onto itself.

OR

Insert the short working end to the right through the loop (just passed through the wraps). Pass the working end around the long standing end and back on itself, forming a bight.

STEP 6. Dress the knot down.

STEP 7. Safety off with an overhand knot against the knot formed on the top single coil. From 10 to 15 centimeters (4 to 6 inches) of rope should be left (standing part).

b. **Checkpoints.**

(1) There are a minimum of four (maximum of six) parallel wraps; the top and bottom ropes cross forward of the hips.

(2) The loop must be underneath all wraps.

(3) Other checkpoints are the same as the bowline.

7-21. THREE-LOOP BOWLINE KNOT

The three-loop bowline knot is used to form three fixed loops in the middle of a rope (Figure 7-23). It is used in a self-equalizing anchor system.

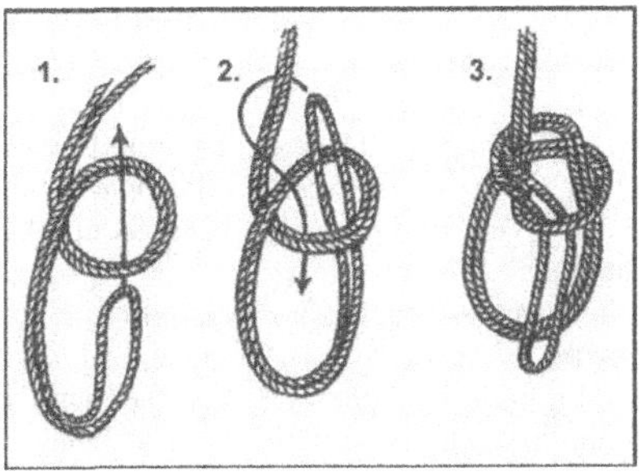

Figure 7-23. Three-loop bowline knot.

a. **Tying the Knot.**

STEP 1. Form about a 60 centimeter (24-inch) bight in the rope, laying it across the left hand, palm up, with the bight to the front.

STEP 2. With the right hand, form a loop with both ropes. Hold it in the left hand.

STEP 3. With the right hand, bring the long bight through the loop(formed in STEP 2) and around and around the two standing ends.

STEP 4. Follow the bight around the standing ends and back down into the loop.

STEP 5. Hold all three loops in the left hand. With the right hand, grasp both standing ropes and pull while dressing up the bowline.

b. **Checkpoints.**

(1) Checkpoints are the same as a regular bowline, except that it is tied with two ropes.

(2) An overhand knot must be tied if the third loop is not to be used as a secondary anchor.

7-22. FIGURE-EIGHT SLIP KNOT

The figure-eight slip knot forms an adjustable bight in a rope (Figure 7-24).

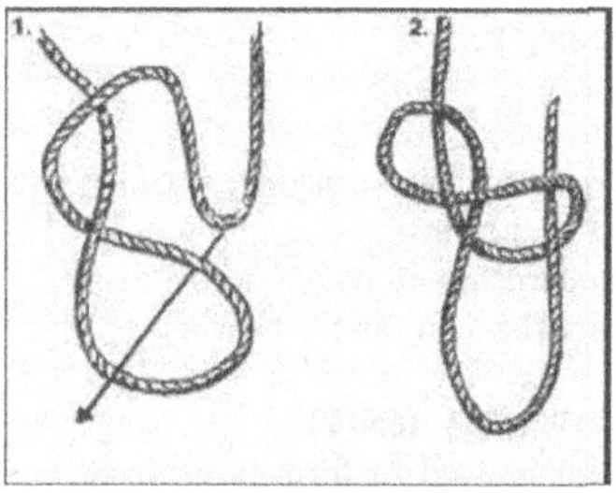

Figure 7-24. Figure-eight slip knot.

a. **Tying the Knot.**

STEP 1. Form a 30-centimeter (12-inch) bight in the end of the rope.

STEP 2. Hold the center of the bight in the right hand. Hold the two parallel ropes from the bight in the left hand about 30 centimeters (12 inches) up the rope.

STEP 3. With the center of the bight in the right hand, twist two complete turns clockwise.

STEP 4. Reach through the bight and grasp the long standing end of the rope. Pull another bight (from the long standing end) back through the original bight.

STEP 5. Pull down on the short working end of the rope and dress the knot down.

STEP 6. If the knot is to be used in a transport tightening system, take the working end of the rope and form a half hitch around the loop of the figure-eight knot.

b. **Checkpoints.**

(1) The knot is in the shape of a figure eight.

(2) The sliding portion of the rope is the long working end of the rope.

Section V. SPECIAL-PURPOSE KNOTS

7-23. KLEIMHIEST KNOT

The kleimhiest knot provides a moveable, easily adjustable, high-tension knot capable of holding extremely heavy loads while being pulled tight (Figure 7-25). This system requires a six-man pulling team when used in a transport tightening system.

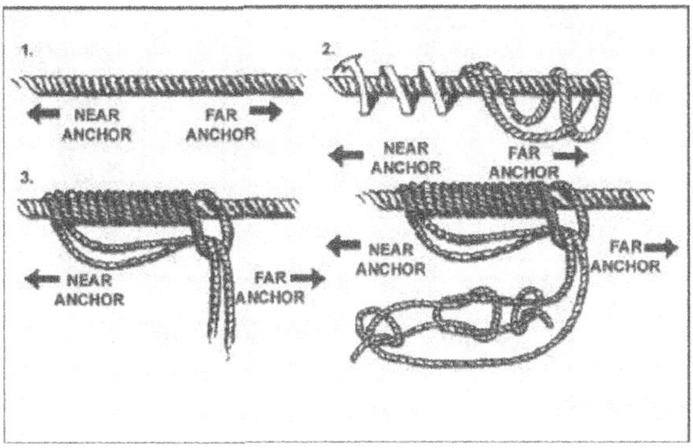

Figure 7-25. Kleimhiest knot.

a. **Tying the Knot.**

STEP 1. Using a utility rope, offset the ends by 30 centimeters (12inches). With the ends offset, find the center of the rope and form a bight. Lay the bight over a horizontal rope near the far side anchor.

STEP 2. Wrap the tails of the utility rope around the horizontal rope back toward the near side anchor. Wrap at least four complete turns.

STEP 3. Pass the remaining tails of the utility rope through the bight (see STEP 1).

STEP 4. Join the two ends of the tail with a joining knot (square knot with overhand safety knots).

STEP 5. Dress the knot down tightly so that all wraps are touching. Insert two snaplinks (opposing gates) into the large loop tied with the tails.

b. **Checkpoints.**

(1) The bight is closest to the far side anchor.

(2) All wraps are tight and touching.

(3) The ends of the utility rope are properly secured with the adjoining knot.

7-24. OVERHAND KNOT

The overhand knot (Figure 7-26) is used to make a knotted rope for a handline, to secure the ends of other knots, and to make stirrups in direct-aid climbing. This knot can also be used to temporarily whip the end of a rope. It is used with other knots.

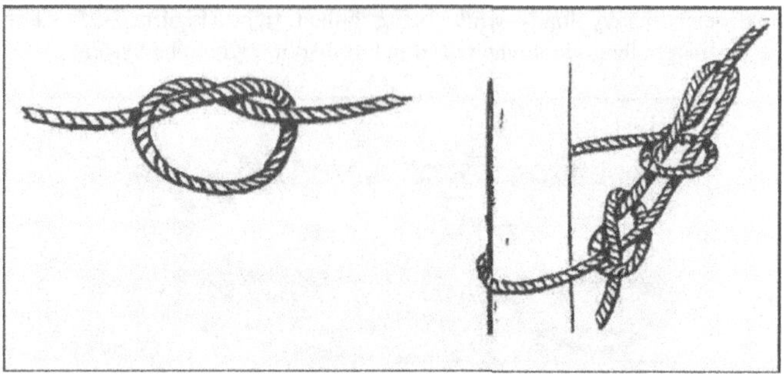

Figure 7-26. Overhand knot.

a. **Tying the Knot.**

STEP 1. Make a loop in the rope.
STEP 2. Pass the working end of the rope through the loop.
STEP 3. Dress down the knot.

b. **Checkpoints.**

(1) Running ends exit the loop in opposite directions.
(2) The knot is dressed down tightly.

7-25. FROST KNOT

The frost knot is used when working with flat webbing (Figure 7-27). It holds better than a standard overhand knot or square knot.

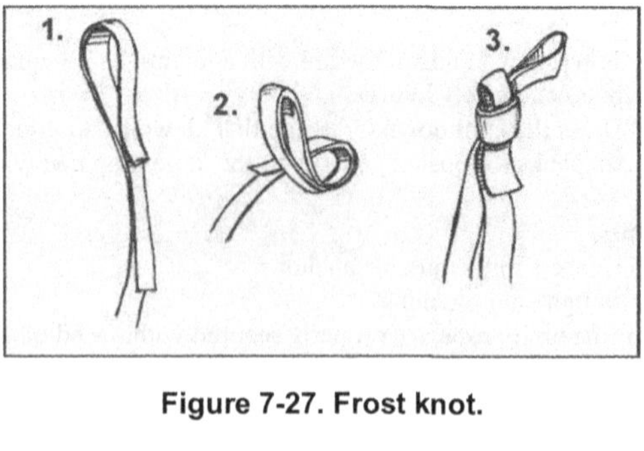

Figure 7-27. Frost knot.

a. **Tying the Knot.**

STEP 1. Lap one end (a bight) of webbing over the other (about 25 to 30 centimeters [10 to 12 inches]).
STEP 2. Tie an overhand knot with the newly formed triple-strand webbing; dress tightly.

b. **Checkpoints.**
(1) The tails of the webbing run in opposite directions.
(2) Three strands of webbing are formed into a tight overhand knot.
(3) A bight and tail exit the top of the overhand knot.

7-26. GIRTH HITCH KNOT
The girth hitch knot is used in tying a runner to a partly driven piton (Figure 7-28).

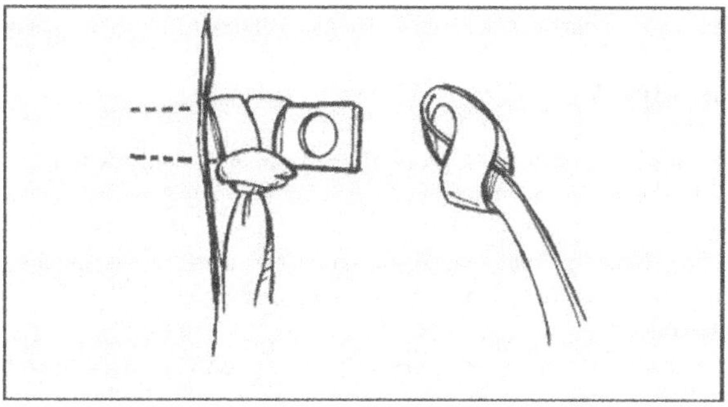

Figure 7-28. Girth hitch knot.

a. **Tying the Knot.**

STEP 1. Bring the standing ends back through the bight of a length of rope or webbing.
STEP 2. Dress the knot tightly.

b. **Checkpoints.**
(1) Two wraps exist with a locking bar running across the wraps.
(2) The knot is dressed tightly.

APPENDIX A

RISK ASSESSMENT

This appendix explains risk assessment for rappel operations. A risk assessement matrix is shown in Table A-1.

A-1. EFFECT

a. **Catastrophic.** Death or permanent total disability, system loss, major property damage.

b. **Critical.** Permanent partial disability, temporary total disability in excess of three months, major system damage, significant property damage.

c. **Moderate.** Minor injury, lost workday accident, compensable injury or illness, minor system damage, minor property damage.

d. **Negligible.** First aid or minor supportive medical treatment, minor system impairment.

A-2. PROBABILITY

a. **Frequent.**

(1) Individual soldier or item—Occurs often in career or equipment service life.

(2) All soldiers exposed or item inventory—Continuously experienced.

b. **Likely.**

(1) Individual soldier or item—Occurs several times in career or equipment service life.

(2) All soldier exposed or item inventory—Occurs frequently.

c. **Occasional.**

(1) Individual soldier or item—Occurs sometime in career or equipment service life.

(2) All soldiers exposed or item inventory—Occurs sporadically, or several times in inventory service life.

d. **Seldom.**

(1) Individual soldier or item—Possible to occur in career or equipment service life.

(2) All soldiers exposed or item inventory—Remote chance of occurrence; expected to occur sometime in inventory service life.

e. **Unlikely.**

(1) Individual soldier or item—Can assume will not occur in career or equipment service life.

(2) All soldiers exposed or item inventory—Possible, but improbable; occurs only very rarely.

A-3. RISK LEVELS

a. **Extremely High.** Loss of ability to accomplish mission.

b. **High.** Significantly degrades mission capabilities in terms of required mission standards.

c. **Medium.** Degrades mission capabilities in terms of required mission standards.

d. **Low.** Little or no impact on mission accomplishment.

RISK ASSESSMENT			HAZARD PROBABILITY				
			Frequent	Likely	Occasional	Seldom	Unlikely
			A	B	C	D	E
EFFECT	Catastrophic	I	***Extremely High***				
	Critical	II			***High***		
	Moderate	III		***Medium***			
	Negligible	IV			***Low***		

Table A-1. Risk assessment matrix.

APPENDIX B

BASIC EQUIPMENT FOR RAPPEL OPERATIONS

The equipment listed in this appendix is the standard basic items required to conduct ground and air rappelling operations.

B-1. GROUND OPERATIONS

For ground operations, the following equipment is needed:

a. Nylon rope, 120 feet (NSN 4020-00-931-8793).
b. Snaplinks (NSN 8465-00-360-0228).
c. Gloves (NSN 8415-00-268-7868).
d. Rope coiling log or standard deployment bag.
e. Rappel seat.
f. Rappel rings (used as expedient anchor points).

B-2. AIR OPERATIONS

For air operations, the following equipment is needed:

a. All items in paragraph B-1.
b. Donut ring (UH-1H operations only) (locally produced TSC item).
c. Safety floating ring.
d. Two 6-foot safety ropes (with snaplinks).
e. Eight 120-foot nylon ropes.
f. Monkey harness for static rappel master.
g. Sixteen snaplinks (UH-60 operations).
h. Twenty-two snaplinks (UH-1H operations).
i. FM communications.
j. Aircraft communications headset.

GLOSSARY

ACRONYMS AND ABBREVIATIONS

AFRM	assistant fast-rope master
AGL	above ground level
AMC	air mission commander
AR	Army regulation
ATP	aircrew training program
attn	attention
BDU	battle dress uniform
CG	commanding general
CGU	modified helicopter safety strap
chemlight	chemical light
cm	centimeter
DA	Department of the Army
D-bag	deployment bag
F	Fahrenheit
FM	frequency modulated; field manual
FRIES	fast-rope insertion/extraction system
FRM	fast-rope master
GSO	ground safety officer
HQ	headquarters
IAW	in accordance with
ICS	interagency communications system
intercomm	intercommunication
LBE	load-bearing equipment
LZ	landing zone
MEDEVAC	medical evacuation
mm	millimeter
mph	miles per hour
MSO	mountaineering safety officer
NCO	noncommissioned officer
NCOIC	noncommissioned officer in charge
NSN	national stock number
NVG	night vision goggles

NWTC	Northern Warfare Training Center
OIC	officer in charge
PIC	pilot in command
POI	program of instruction
POL	petroleum, oils, and lubricants
RATELO	radiotelephone operator
RSO	rappel safety officer
SFC	sergeant first class
SOAR	Special Operations Aviation Regiment
SOP	standing operating procedures
SPIES	special patrol infiltration/exfiltration system
STABO	a system for extracting personnel by helicopter (the combined first letters of the surnames of the five persons who designed the system)
SSO	STABO system operator
TAPO	Technology Applications Program Office
TC	training circular
TM	technical manual
TRADOC	Training Doctrine Command
TSC	training support center
UIAA	Union of International Alpine Association
UH	utility helicopter
US	United States
USAF	United States Air Force
USAIS	United States Army Infantry School
USASOC	United States Army Special Operations Command
VIP	very important person

REFERENCES

DOCUMENTS NEEDED

These documents must be available to the intended users of this publication.

DA Form 2028	Recommended Changes to Publications and Blank Forms. 1 February 1974.
DA Form 5752-R	Rope History and Usage. May 1989.
FM 57-38*	Pathfinder Operations. 9 April 1993.
FM 7-93*	Long Range Surveillance Unit Operations. 3 October 1995.
TC 90-6-1*	Military Mountaineering. 26 April 1989.

READINGS RECOMMENDED

These sources contain relevant supplemental information.

AR 95-3	Aviation: General Provisions, Training, and Standardization, A Resource Management. 27 September 1990.
FM 8-35	Evacuation of the Sick and Wounded. 22 December 1983
FM 57-220	Basic Parachuting Techniques and Training. 1 June 1990.
TM 10-1670-262-12	Operator's and Organizational Maintenance Manual (Including Repair Parts and Special Tools List) for Personnel STABO Extraction System and Cargo Personnel Lowering Anchor Device. 7 April 1988.
USASOC Reg 350-2	Training Airborne Operations. 1 June 1995.
USASOC Reg 350-6	Training Special Operations Forces Infiltration/Exfiltration Techniques. 12 May 1995.
	The Air Assault School Handbook.

*This source was also used to develop this publication.

INDEX

TC 21-24
10 SEPTEMBER 1997

By Order of the Secretary of the Army:

DENNIS J. REIMER
General, United States Army
Chief of Staff

Official:

JOEL B. HUDSON
Administrative Assistant to the
Secretary of the Army
03807

DISTRIBUTION:

Active Army, Army National Guard, and U.S. Army Reserve: To be distributed in accordance with the initial distribution number 115059, requirements for TC 21-24.

Made in United States
North Haven, CT
04 November 2026

21774384R00143